Television In The Cloud

Frank A. Aycock, Ph.D.
Adam Powers

Copyright ©2014 Frank Aycock
All rights reserved.

ISBN-13: 978-1494287085
ISBN-10: 1494287080

To my wonderful wife, Gail,
My two sons, Allen and Wes,
And my daughter-in-law, Katie
                                      Frank A. Aycock

For my wife Sara in appreciation of her support,
And for my children, Julia and Miles, may they
learn to read soon
                                      Adam Powers

TABLE OF CONTENTS

ACKNOWLEDGEMENTS vii
PREFACE 9

PART I: THE CLOUD

Chapter 1: Introduction To The Cloud 15
Chapter 2: Types Of Clouds 25
Chapter 3: Cloud Service Models 39
Chapter 4: Advantages Of The Cloud 49

PART II: TELEVISION IN THE CLOUD

Chapter 5: Building Blocks of Television In The Cloud 61
Chapter 6: Moving Television To The Cloud 73
Chapter 7: The Cloud And Television 87
Chapter 8: Making Use Of Television In The Cloud 99
Chapter 9: The Freedoms Of Television In The Cloud 113
Chapter 10: The Future Of Television In The Cloud 131

NOTES 142
INDEX 155

ACKNOWLEDGEMENTS

Writing a book is never easy. However, the endeavor is made much easier if you have a great team working with you. It has been a pleasure having the opportunity to put together a great team of professionals to assist me in writing this book.

First, my thanks go out to Adam Powers, my co-author and Vice President Media Technology & Solutions at V2 Solutions. Adam was kind enough to come on board with me on this project at the last minute and has been invaluable in making this book a success. Without Adam, this book most likely could not have been finished as easily as it has been.

I want to thank my colleague and friend Dr. Larry Taylor for his assistance in designing the cover of the book. Dr. Taylor has been kind enough to collaborate with me on both my books, and his creativity has been a real blessing.

I also want to thank my colleague, Roger Gonce, for the layout of the book. Like Adam, Roger came to the project at a late point, and rescued me at the last minute. His expertise and enthusiasm for the project has also been a real blessing and made it possible for the book to be finished in a timely manner.

There are times when a person offers you an opportunity to participate in an endeavor that changes the direction of your life – if not permanently, then on a temporary basis. Marty Lafferty, the Chief Executive Officer of the DCIA, the Distributed Computing Industry Association, offered me the opportunity to participate in the NAB's cloud conference in 2013. Marty's willingness to add me to his roster of amazing cloud professionals for the conference is the real genesis of this book. I was already working on a different book project, but Marty's enthusiasm and support along the way has been the encouragement I needed to bring this endeavor to fruition.

I would like to thank Michael Rofe, global chairman of the iCTVBA, for his support and his enthusiasm for this project. Michael has been a staunch supporter of my work since my first book, and a real friend to my wife and me.

Finally, and most importantly, I want to thank Gail, my beautiful and loving wife of 33 years, without whom I know I would still be slowly working on this book. She is my support, my cheerleader, and my copyeditor. She is always right beside me every step of the way, and she believes in every book I write and in me. I am truly a blessed man.

Frank A. Aycock
February 2014

PREFACE

Television In The Cloud is the second installment in the book series *21st Century Television*. The initial volume, *21st Century Television: The Players, The Viewers, The Money*, provided an overview of the coming television revolution. The volume discussed in detail the major players in the television industry – broadly defined – looking first at the legacy industries of the broadcasters, the cable industry, and the direct-to-home satellite industry. The book then discussed the new media players – Internet-protocol television, video-on-demand, the over-the-top set top box, mobile DTV, connected (or "smart") TVs, and the iWorld of smartphones, tablets, and apps. The first section then ended with a discussion and general overview of the cloud and its relationship to television.

Following the discussion of the major players, the book discussed the changing television viewer, from the members of the "greatest generation" to the youngest of generations, the "TV Next-Gen" generation. The section looked at the changing media habits and the changing uses of television.

Finally, the book looked at monetizing 21st Century Television, including new ways of advertising, the development of ubiquitous product placement, retransmission consent fees, and new promotional developments. The section ended with a look at the development of television as a global entity instead of a national, regional, or local entity. The book was controversial in that it provided, as one reader put it "a blueprint for the future of television."

PREFACE

Television In The Cloud examines in great depth what the chapter on the cloud in the previous book only looked at in a more superficial way. While the original cloud chapter provided a basic overview of the cloud and its relationship to television, *Television In The Cloud* delves deeply into the cloud technology – what it is, how it operates, and how it will change television forever. You see, television is made for the cloud, and the cloud is made for television. It's a phrase the reader will see repeated time and again throughout this book, because it is the core underpinning of 21st Century Television. Television and the cloud – when brought together in their unique bonding ability – change the entire way of producing and distributing television and make possible the beauty that is the 21st Century Television universe.

Television In The Cloud has been written for the television professional who finds him/herself contemplating, considering, or even wondering if a move to the cloud is right for his/her television organization. This book is emphatically NOT a book for the television engineer who is already involved in cloud computing. The book is way too plain-spoken for the engineer to find interesting, most likely, *unless* s/he has been asked/ordered for the first time to find out what cloud television is and whether it is right for her/his television company. However, for those professionals that are contemplating or just beginning to make their first moves into this exciting and enticing new realm of 21st Century Television known as Television In The Cloud, this book can be a guide in making decisions on whether and how to enter the world of Television In The Cloud.

To truly understand Television In The Cloud, the reader must first understand cloud computing. Therefore, the first section of the book will provide in-depth discussions of what the cloud is, the different types of clouds that are available, and the various types of service models that exist. Written in a plain, straightforward manner, the section takes the somewhat-but-not-overly technical topics and brings them into language that the professional can easily understand. Understanding the first section provides a solid foundation for the rest of *Television In The Cloud*.

Following the basics of cloud computing, the book moves into the most technical of the chapters. These two chapters will likely be the most enjoyable for engineers due to their more technical nature. The chapters will look at the building blocks of cloud television and the process of delivering television in the cloud from source (the content provider) to end user (the television viewer). While realizing the much more highly technical nature of these chapters, your authors have worked diligently to try as much as possible to keep the discussions of these chapters in the same plain, straightforward language that is in keeping with the rest of the book.

The final chapters of the book are devoted to the ways in which Television In The Cloud can transform the areas of 21st Century Television, from production, to programming, to news, to sales, etc. Each aspect of what is considered television today is discussed in-depth in light of the transformative power of the cloud. In transforming every aspect of today's television, Television In The Cloud will move today's television into that magical world known as the 21st Century Television universe. In doing so, T*elevision In The Cloud* will tie together the new and profoundly topical ideas discussed in *21st Century Television: The Players, The Viewers, The Money* into a revolutionary whole. Finally, the book discusses in detail the benefits derived from moving television into the cloud, and finishes with a discussion of Television In The Cloud as a liberating source of freedom for the television viewers and the television industries alike around the world. Remember: Television is made for the cloud and the cloud is made for television. Enjoy the revolution!

PART I
THE CLOUD

CHAPTER 1
INTRODUCTION TO THE CLOUD

Before it is possible to discuss Television In The Cloud, it is important to have a fuller understanding of what the cloud is, what types of clouds there are, what the benefits of the cloud are, and cloud service models. As such, this book will look at the broader picture of cloud computing before going into the narrower topic of Television In The Cloud. Consider Television In The Cloud as a subset of the larger cloud computing.

Now, for those persons who, when the term "the cloud" is mentioned, immediately thinks, "Cirrus, Cumulus. . .," this chapter and the ones that follow may be real eye-openers. While the term that's used, and will be used throughout this book, is "the cloud," a phrase stated in the singular, in reality, the more correct term should be "the clouds" – plural – because there are many clouds in the digital world and have been since before the term "the cloud" was coined for everyday use.

To understand where Television In The Cloud is heading, it is important first to define the terms "cloud computing" and "the cloud." Both terms can be discussed from the more formal viewpoints of dictionaries as well as the more professionally-oriented viewpoints coming from today's trade literature. First, let's look at what the more formal definitions are.

The *Oxford Dictionary* defines cloud computing as, "the practice of using a network of remote servers hosted to the Internet

THE CLOUD

to store, manage, and process data, rather than a local server or a personal computer."[1] *Dictionary.com's 21st Century Lexicon* defines "cloud computing" as, "a type of computing based on sharing computing resources rather than having local servers or personal devices to handle applications."[2] *The Tech Terms Computer Dictionary* defines cloud computing as "applications and services offered over the Internet. These services are offered from data centers all over the world, which collectively are referred to as the 'cloud.'"[3] Chris Poelker of *Computerworld* defines cloud computing as "simply a way to describe how organizations can take some or all of their existing IT infrastructure and operations and hand it over to someone else to build or manage, so the internal core team can focus on new ways to help the business at hand...."[4] Finally, the National Institute of Standards and Technology, a part of the U.S. Department of Commerce, provides what is the official U.S. government's definition of cloud computing in its Special Publication 800-145 *The NIST Definition of Cloud Computing*: "Cloud computing is a model for enabling ubiquitous, convenient, on-demand network access to a shared pool of configurable computing resources (e.g., networks, servers, storage, applications, and services) that can be rapidly provisioned and released with minimal management effort or service provider interaction. This cloud model is composed of five essential characteristics, three service models, and four deployment models."[5] Unless you are an engineer, or a true computer geek, this last definition probably leaves you cold. After all, it is government work. However, it is probably one of the, if not the, most precise definition of cloud computing available.

In any attempt to define cloud computing one must also look only at the term, "cloud" and try to define it as well. One definition of the cloud is "a set of hardware, networks, storage, services, and interfaces that enable the delivery of computing as a service."[6] An even broader (and perhaps a better) definition for the term is that the cloud is simply "a metaphor for the Internet (based on how it is depicted in computer network diagrams) and is an abstraction for the complex infrastructure it conceals."[7]

Looking at the various definitions of cloud and cloud computing above, there are several important aspects to the cloud. First, a cloud must be tied to the Internet. Every one of the definitions of cloud computing states specifically or suggests that cloud computing – and for that matter, the cloud itself - must make use of the Internet for its operations. It is the Internet that provides the cloud with the ability to be accessed from anywhere at any time by any platform capable of accessing the Internet. One of the definitions of cloud even defines it as a metaphor for the Internet itself, a good way of considering the cloud these days.

Second, while **the** cloud at its ultimate may be metaphorical for the Internet, **a** cloud is a subset of the Internet. In other words, there are many clouds that are part of the largest cloud, the Internet itself. Depending on whose work is considered, a cloud can be categorized into one of two, three, or four groups. At the broadest of the categorizations (4 groups), they are public, private, community, and hybrid.[8] Each of these types of clouds will be discussed in depth in Chapter 2. Each category of clouds can be specific for a small subset of individuals (private clouds) all the way to massive numbers of individuals (public clouds). However, it is the use and structure of the cloud that determines into which category a cloud will fit.

Third, a cloud includes a specific group that can access that particular cloud, but is not available for everyone to access. While the Internet can be accessed by anyone who has the appropriate technology, the user of a particular cloud has to have some individual way of connecting to that cloud that others outside of that particular cloud do not have. This is especially true of private clouds, community clouds, and hybrid clouds, where security ranges from extremely important to paramount, and can be a constant concern for cloud users. Security, in such cases, can refer both to protection from intrusion by outsiders and protection from loss of stored information. Chapter 4 will look more deeply at the advantages of using the cloud.

Fourth, the source of a cloud generally is not local. In other words, it is not internally connected within a company, but rather connected externally through the Internet. Even a private cloud, one that is maintained within the confines of an organization, must still be connected to the Internet so that the employees can have access anytime, anywhere, and on any platform. Today, however, there is an explosion of cloud services that makes it possible for companies to store, access, and retrieve their data without the necessity of having the necessary racks and racks of servers at the location. Amazon's AWS service, Rackspace, and Microsoft's Azure service are just three of the many examples of companies offering cloud solutions for a fee.

Fifth, for the specific group of people who are allowed access to a particular cloud, they can access that cloud from anywhere they can get an Internet connection. That access, then, is not limited, except where governments deny the possibility to connect to the Internet. A businessperson from Florida with access to a particular cloud hosted in California can connect with that cloud even if (s)he is in France, Japan, Australia, or anywhere else in the world that the businessperson can connect with the Internet.

For television, this last point is extremely important. The ability of the cloud to allow the viewer the opportunity to enjoy his/her favorite program no matter where (s)he is in the world, is a major driver of the future of television. In the future, using Television In The Cloud technology, content providers will be able to offer their programing across the globe to audiences on any continent or island, providing there is Internet connectivity. In addition, advertisers will be able to deliver customized, addressable advertising to viewers no matter where they are in the world. Further, advertisers will be able to reach potential consumers in areas of the world that they may not have been able to reach previously, in a cost-effective manner. This ability is due to the programming on which they place their advertising being able to reach any and all locations across the globe. Finally, television viewers around the world will have the freedom to enjoy programming from content providers everywhere, making the first two points in this paragraph financially advantageous.[9]

Despite the current definitions of the terms "cloud" and "cloud computing," in reality, precursor forms of the cloud have been around for a while. Using the definitions in the earlier paragraph, there are two examples of what can reasonably be considered the forerunners of today's clouds.

The first of these is the use of servers to tie together all the computers in an organization. While these computers are connected together, it is through an *intranet* rather than through the *Internet*. While these connections may have some similarities to a cloud, looking at the five requirements above, these intranets, depending on how they are constructed, lack most or all of the requirements for being a cloud. As an example, my (Aycock) office at the university is connected to a series of servers that the university owns and operates. On one or more of those servers are located the software that I use every day, and the different work files that I produce. I do not have to use a browser to connect and produce my work, but, at the same time, the software and work files are not on my specific computer. Further, until recently, I was not able to work on my files at home or on the road unless I downloaded them, and I could not upload the files back to the server without being in my office. This is an example of an intranet but not a cloud. While this example does meet the third requirement listed above and could be considered the exception to the fourth requirement much as a private cloud could be, the intranet does not meet the first, second, and fifth requirements. Today, however, because the university has created its own private cloud system, the university has made it possible for me to access my work from anywhere using its cloud.

The second example is the home that has several computers tied to a specific server. Much like the first example above, this example is, technically, not a cloud for the same reasons as the previous example. While a homeowner might say that (s)he has a private cloud with this arrangement, in reality, it is not a true cloud. What the homeowner actually has developed is a home intranet.

History of Cloud Computing

While the first actual cloud service did not come along until the late-1990s, the precursor theoretical underpinnings of cloud computing are generally considered to have been put forth in the 1960s. However, as a precursor theory to underpin the ultimate development of the cloud and cloud computing, a good place to start is a 1945 article by Vannevar Bush in *The Atlantic* magazine entitled "As We May Think." Bush has been described as "the pivotal figure in hypertext research," and directly inspired such notable persons as Douglas Engelbart and Ted Nelson.[10] In his article, Bush put forth the notion of associative thinking, that the information in the world at that time (remember this is 1945!) and what would come in the future was too vast and too complex to find and use through linear thinking. Instead, he suggested that all information should be tied together in much the same way as the human mind works. Bush suggested that, "When data of any sort are placed in storage they are filed alphabetically or numerically, and information is found (when it is) by tracing it down from subclass to subclass. …Having found one item, moreover, one has to emerge from the system and re-enter on a new path."[11]

Instead of linear application for finding data, Bush suggested that there should be a way to organize data so that retrieval occurred much as the mind works. Bush said that the human mind "operates by association. With one item in its grasp, it snaps instantly to the next that is suggested by the association of thought, in accordance with some intricate web of trails carried by the cells of the brain."[12] To demonstrate his ideas, Bush proposed a theoretical "memex" machine and explained how it would accomplish this associated search and retrieval of information. Bush's work inspired others which would lead us to where we are today. While his theorized memex machine was mechanical and – by today's standards – rather quaint, his notions of associative storage and retrieval of data can be considered the underlying foundation on which today's cloud computing, the Internet, and the World Wide Web are based.

It is in the 1960s that many researchers credit the beginnings of the concept of what would become cloud computing. One of the earliest of these pioneers was Joseph Carl Robnett (J.C.R.) Licklider. Licklider is considered to be one of the U.S.'s most important computer scientists and is credited with being the driving force behind ARPANET, the forerunner of the Internet.[13,14] Licklider's vision for the future was a world interconnected with the ability to access any program anywhere, anytime. Margaret Lewis, product marketing director for AMD says, "It is a vision that sounds a lot like what we are calling cloud computing."[15]

Also credited by researchers with the earliest conception of what would become cloud computing was John McCarthy (famous for coining the term "artificial intelligence"), who suggested that computation should be organized along the lines of a public utility.[16,17] Both men's visions of a worldwide computer network would come about. Bush's, Licklider's, and McCarthy's visions would pioneer the way to today's cloud computing and establish a framework to deliver tomorrow's work in the cloud.

While each of these men described a vision of cloud computing, it wasn't until the 1990's that cloud computing as a term came into existence. Dr. Ramnath Chellapa,[18] is credited with the first use of the term cloud computing in his talk entitled *Intermediaries in Cloud-Computing: A New Computing Paradigm*, presented at the INFORMS meeting in Dallas in 1997.[19] Two years later, the first cloud computing company, Salesforce.com, pioneered the use of cloud computing to deliver enterprise applications using their website. Salesforce.com's use of cloud computing opened the door for software companies to begin delivering services over the web.[20]

Three years later, in 2002, Amazon launched its Amazon Web Services, better known today as AWS, the first major suite of cloud-based services. The company followed its initial AWS product in 2006 by introducing its Elastic compute cloud (EC2) to the world. This suite of services made it possible for small companies and even individuals to use the cloud for business and/or personal purposes at affordable prices. It was also designed to make it easier for those same businesses and individuals to use the service.[21] According to Jeremy Allaire, CEO of Brightcove, "Amazon EC2/S3 was the first widely accessible cloud computing infrastructure service."[22]

It was also in 2006 that Google became a major player in cloud computing with the company's introduction of Google Docs, a free-to-use, cloud-based suite of services that brought the notion of cloud computing to the general public.[23] Chris Anderson, author of the book *The Long Tail*, also helped to make Google Docs well known through his description of writing his book *FREE!: The Future of a Radical Price* on Google Docs while sitting at Starbucks.[24]

By 2008, the first platform designed to provide the ability of organizations to operate private clouds was introduced. The Eucalyptus platform is an open-source API-compatible platform that is compatible with Amazon Web Services and is a part of the Amazon Partner Network.[25] By 2010, cloud computing was fully underway with Amazon and Google offering their different suites of programs; Eucalyptus was available for private cloud development; and Microsoft introduced its own cloud platform, Azure.[26] Furthermore, that same year, Rackspace – which had actually been founded in 1998 and focused on hybrid cloud platforms[27] - introduced Openstack[28] which is a global collaboration of developers and cloud computing technologists that are producing a different open source cloud platform from Eucalyptus.

Television In The Cloud

Television is made for the cloud and the cloud is made for television. There's no doubt about the future of television being in the cloud. The opportunities and benefits the cloud affords are too numerous to overlook.[29] Virtually all aspects of television can benefit from what the cloud offers.

Today there are numerous companies that recognize the importance that the cloud will play in the development of 21st Century Television, with more companies offering their products every day it seems. Some companies are specific to what they do; others such as Amazon's AWS, Google's offerings, Rackspace, and Microsoft provide more generalized services that can be tailored to the needs of individual television outlets. Some provide the ability of private clouds; others provide the ability to have a location on a third-party cloud. Regardless, there is a company – or soon will be – that can cater to the needs of any television outlet, broadly defined.

However, Television In The Cloud is still in its infancy. Expanded bandwidth, upload, and especially download speeds must increase significantly to make it possible for high definition, and, in the very near future, ultra-high definition (known as UHD or 4K) television programming to be delivered to the television viewer without buffering or loss of signal. Farther on the horizon will be super-high definition (also called 8K) television and even holographic television, both of which will require even more bandwidth and faster upload/download speeds. Legacy television media outlets have been slow to participate in the development of Television In The Cloud by moving to some form of Internet-delivered television (most likely IPTV). The engineering requirements necessary for real-time delivery of television that is safe, secure, and affordable to the consumer are still in development. As with any new technological advance there are those who eagerly embrace it while others remain skeptical.[30]

Nevertheless, given the huge amounts of money that companies are directing into the development of cloud services for the various television technologies, it seems that Television In The Cloud will become a certainty very shortly and that the sons and daughters of the millennial generation will consider watching television programming from the cloud just an everyday occurrence. At that point, the entire range of possibilities that is the promise of 21st Century Television will be well on its way to becoming reality.[31] Television is, truly, made for the cloud, and the cloud is, truly, made for television!

CHAPTER 2
TYPES OF CLOUDS

Most often known as deployment models, there is more than one type of cloud. However, depending on the article or book that one reads, the number of cloud types discussed may be different. Some articles and books only refer to two cloud types – public and private. Others will discuss three types of clouds – public, private, and hybrid. Still others will identify four different cloud types – public, private, community, and hybrid. This chapter will discuss the four different cloud types, for the main reasons that (1) discussing the four types gives more understanding of the differences in cloud possibilities, and (2) because there are four major types of clouds in use today as identified by the National Institute for Standards and Technology of the U.S. Department of Commerce. Those four types of clouds – mentioned in the previous chapter – are public clouds, community clouds, hybrid clouds, and private clouds.[1] Of the four types of clouds listed, the one that is most questionable is the community cloud. Depending on who or what one references, the community cloud may be a separate category of clouds, or it may simply be a subset of the private cloud. Both aspects will be discussed in this chapter.

Each of these types of clouds is designed for specific purposes, and with each one, the sphere of users generally becomes smaller. However, regardless of the cloud being discussed, each of the four will conform to the five requirements of a cloud discussed earlier in Chapter 1. As a reminder, those five requirements are:

1. The cloud – whether a public, community, hybrid, or private cloud - must be tied to the Internet, which is the ultimate cloud.

2. The cloud must be a subset of the Internet. How large the subset depends on the type and usage of the cloud. For some public clouds, that subset can be – relatively speaking – exceedingly large. For private clouds, on the other hand, that subset can be – again, relatively speaking – very small, encompassing only a limited number of users.

3. The cloud must include a specific group that can access that particular cloud, but is not available for everyone to access. Even the largest of public clouds, will, in some way, be limiting in the availability of usage.

4. With very specific exceptions, the source of a cloud generally will not be local. This requirement depends on the type and usage of the cloud. For instance, a private cloud may be designed for the use of employees in a company. The company may choose to build and maintain its cloud onsite. However, in most instances, clouds are located offsite, and often are run by third-party operators.

5. For the specific group of people who are allowed access to a particular cloud, they must be able to access that cloud from anywhere they can get Internet access. This requirement is key for all clouds. For anyone who has the right to access a particular cloud, regardless of the type of cloud, that person must have the ability to access the cloud from anywhere that (s)he is able to connect to the Internet. It is this requirement that will make the cloud the underpinning for 21st Century Television.

The Public Cloud

The public cloud is one of the latest innovations in the overall evolution of computing. Public clouds are designed and implemented by third-party providers and are located off-site from their clients. The public cloud is generally made available to all users interested in making use of the public cloud offered by the third-party service provider. Some of these cloud services are available free of charge; others are designed as pay-per-use, whereby a client can purchase as much of the cloud as is needed to accomplish the goals of the organization. Additionally, when a public cloud operates on a pay-as-you-go basis, the client can add or reduce the amount of cloud purchased, depending on the need of the organization at any given time. Further, just like the clients that a cloud provider may have, the provider can also add additional racks of servers to increase the amount of storage and/or operational capabilities of the provider's cloud, as well as to expand the provider's cloud service offerings. Examples of public clouds include Google Docs, Apple's iCloud, or Amazon's cloud. The public is free to make use of Google Docs, free to join Apple's iCloud, and free to join Amazon's cloud as well. While a person may have to buy a song or a book to have a use for Apple's and Amazon's cloud services, there are no limitations on who can be a part of either cloud service.

In his whitepaper sponsored by Trend Micro, Inc., analyst and researcher Krishnan Subramanian describes what he refers to as "unique advantages" that public clouds have.[2] Subramanian says, "The use of public cloud services shifts the responsibility of managing complex IT, which is not the core business of many companies, to a third-part provider, thereby, offering some benefits that cannot be realized either in traditional infrastructure or private clouds."[3] He describes two major benefits offered by public clouds:

Cost Savings:
- Eliminates capex [capital expenditures] and offers reduced opex (operating expenditures) because the maintenance and labor costs associated with managing the infrastructure is offloaded to a third-party provider.

- Ensures cost efficiency because of the pay-per-use models. Typically, service providers charge by the hour and this comes in handy when a company's resource needs are for a temporary project or to meet a sudden spike in usage, avoiding the need to build out internal infrastructure to cover these projects.

- Offers self-service provisioning, leading to lower costs and better agility because human intervention in resource provisioning is minimized.[4]

Business Agility:

• Provides massive scalability and an ability to elastically re-size compute resources based on the organization's IT needs.

• Gives programmatic access to computer resources through API, helping applications scale automatically without any human intervention.

• Supplies robust infrastructure with better support staff, offering cost and talent advantages in an increasingly shrinking pool of expertise.[5]

Even though the public cloud offers a number of advantages, it does have a very glaring weakness – that of security. Any business determining to use a public cloud must weigh the security risks that are inherent in public clouds. Charlie Williams, marketing director for 2x.com, a cloud computing software company[6] (as well as others[7]), identifies four major security concerns for firms using public clouds.

First, there is the multi-tenancy concern. Public clouds are, by nature – well, public – which means that there are numerous companies of varying sizes sharing resources. All of these companies can be unrelated to each other, but at the same time they are sharing the same resources, including CPU resources, memory resources, even sharing the same building with the other company-tenants. The problems of multi-tenancy can lead to tenants being able to see data belonging to others, manipulate others' data, or even assume another's identity.[8]

Second, there is the concern over limited control on running technologies. When a company becomes a client of a public cloud vendor, the company is left with little to no control over the business processes that are operating in the cloud. This lack of control can lead to virtual exploits regarding what's happening virtually among the host and cloud clients. Further, without investigating, a client company likely will have no idea what version of the virtualization software is running and likely not even know what virtualization software is even being used. This lack of transparency is not well understood and needs to be known before entering into a public cloud.[9]

Third, there is always the concern with the quality of the encryption used by the cloud vendor. Data protection is crucial for companies – media or otherwise – and how to protect those data from others is at the forefront of today's discussions regarding cloud technology and especially public clouds. How the data are encrypted, the quality of protection the encryption provides, the existence of private keys and who has access to those keys are all critical to having a successful experience with a public cloud. Additionally, though, it is important to know what the retention policies of the cloud vendor are. When the client deletes data from the cloud, it should be erased completely. If the vendor doesn't provide for complete expungement of the data, the client should be made aware beforehand as it opens the client's deleted data to access and use by others.[10]

Williams' final concern is for data compliance. Public clouds store data at random locations, with later versions of the data possibly residing in physical localities that are great distances from one another. The client may have no idea where the data reside and what might be the dangers inherent in the location where the data are, especially if the data are located in a different country. The laws and regulations governing storage, retrieval, encryption, and access could be vastly different from what the client might reasonably expect and could be of enormous concern to the client.[11]

To Williams' four concerns, security analyst Roger Grimes adds a fifth – that of ownership. According to Grimes, many of the cloud vendors, including the largest and most well-known "have clauses in their contracts that explicitly states [sic] that the data stored is the provider's – not the customer's."[12] Grimes goes on to say that the cloud vendors require their joint ownership with the client because "it gives them more legal protections if something goes wrong. Plus, they can search and mine customer data to create additional revenue opportunities for themselves."[13] Such requirements can be very uncomfortable to a cloud client considering using a public cloud.

Nevertheless, public cloud usage is growing, despite the security concerns. The agility of the public cloud, the scalability, the pay-as-you-go option, and the speed of development makes the public cloud the "dominant cloud by far," and one which "has nowhere to go but up."[14]

The Private Cloud

At the opposite end of the cloud spectrum is the private cloud. Whereas the public cloud is available to a wide range of users, the private cloud is designed and implemented for exclusive use by a single organization. A private cloud provides the company with not only a distinct cloud-based environment, but also one that is more secure than the public cloud. The private cloud's resources are only accessible by the members of the organization (employees and management), thus providing the protection often needed by organizations in their businesses. The private cloud is the most restrictive of the clouds, being available only for those connected with the organization that implemented the cloud.

Private clouds are a very popular choice for companies moving into the cloud for the first time, because of the security protections that are inherent in the private cloud. However, the percentage of those first-time cloud users preferring private clouds over public or hybrid clouds is shrinking. In 2011, Beth Schultz of Network World reported that "a recent Info-Tech survey shows that 76% of IT decision-makers will focus initially, or, in the case of 33% of respondents, exclusively on the private cloud."[15] In that same article, Joe Coyle, CTO of Capgemini stated, "The bulk of our clients come in thinking private. They want to understand the cloud, and think it's best to get their feet wet within their own four walls."[16] But, Coyle adds, "Private cloud oftentimes is the knee-jerk reaction, but not necessarily the right decision. What companies really need to do is look at each workload to determine which kind of cloud it should be in."[17] However, as knowledge of the different types of clouds have become more widespread, it seems as if Coyle's suggestion has proven correct. In a Cisco/PC Connection survey of cloud computing, the percentage of organizations contemplating an initial move to a private cloud had dropped to 55%, still more than half, but far below the 76% reported in the previous 2011 Info-Tech survey.[18]

The reason more companies choose the private cloud initially is because the private cloud is closer to the traditional IT infrastructure of a local area network plus virtualization advantages added onto the network. As such, the private cloud has certain benefits that make it comfortable to those organizations. According to the article "What is a Private Cloud," those benefits are:

• Higher security and privacy: ...private clouds – using techniques such as distinct pools of resources with access restricted to connections made from behind one organization's firewall, dedicated leased lines and/or on-site internal hosting – can ensure that operations are kept out of reach of prying eyes,

• More control: as a private cloud is only accessible by a single organization, the organization will have the ability to configure and manage it inline with their needs to achieve a tailored network solution. However, this level of control removes somes [sic] (some of) the economies of scale generated in public clouds by having centralized management of the hardware,

• Cost and energy efficiency: implementing a private cloud model can improve the allocation of resources with an organization by ensuring that the availability of resources to individual departments/business functions can directly and flexibly respond to their demands. ...they do make more efficient use of the computing resource than traditional LANs as they minimize the investment in unused capacity.

• Improved reliability: even where resources (servers, networks, etc.) are hosted internally, the creation of virtualized operating environments means that the network is more resilient to individual failures across the physical infrastructure. Virtual partitions can, for example, pull their resources from the remaining unaffected servers.

• Cloud bursting: some providers may offer the opportunity to employ cloud bursting, within a private cloud offering, in the event of spikes in demand.[19]

The private cloud can be owned, managed, and operated by the organization itself if it chooses to, but can also be owned, managed, and operated by a third-party provider, or by a combination of the organization and the third-party provider jointly. A self-run or stand-alone private cloud is one that is located at the organization's facilities. It is built by the organization and run by it as well. Generally, these types of private clouds are designed to support a limited number of applications that the organization uses in the business.[20] Because the cloud is controlled by the organization, security, flexibility, and control are easy to maintain. However, elasticity, or the ability to "scale up" or enlarge the storage is more difficult to achieve, which leaves the organization operating its cloud at maximum or near-maximum capacity at all times. Additionally, the organization's IT department will be responsible for maintaining the cloud environment which can be costly and take resources away from other areas. Finally, the resilience of the cloud environment must be maintained and the mobility capabilities must be developed internally. The self-run or standalone private cloud is costly and complex to build out and to maintain. The upside is the control and heightened security protections that it offers.[21]

The second type of private cloud deployment model is the managed private cloud. An example of a managed private cloud is a company having its own private cloud, but locating it with a third-party provider. In this type of private cloud, the third-party provider is in charge of supplying the facilities, environmental controls, and connectivity for the organization's network. The organization owns, manages, and controls the cloud environment just as if it was located within the organization's facilities.

Because the organization owns the private cloud despite its location with a third-party provider, much like the self-run private cloud, control, security, and flexibility are easy to maintain. More difficult to maintain is the management of the cloud and the resilience of it. Because responsibility is shared between the organization owning the cloud and the third-party provider, the management abilities are shared as well, with the provider supplying the physical aspects of the cloud (the location, the power to run the facility, and the connectivity from the cloud to the organization for use). Because the third-party provider supplies the physical attributes, the cloud owner is forced to relinquish the management of those aspects. Likewise, it is the third-party provider that supplies the ability to keep the cloud in operation. This resilience – or lack of – is critical to the cloud owner organization, and the working relationship with the third-party provider in this area is crucial. The most difficult aspects to maintain are the elasticity (or scalability) and the ability to develop and maintain a solid mobile infrastructure. Both aspects are challenging for the same reason – the cloud the organization is using is housed off-site, while the responsibility for both aspects belongs to the organization, not the third-party provider. Further, in both cases, because of the previous reasons, both aspects will incur delays not found in the self-run private cloud.[22]

Finally, there is the dedicated private cloud deployment model. In this deployment model, the organization owns the cloud, but the third-party provider both operates and maintains the cloud for the organization. Generally speaking, the third-party provider is an experienced cloud provider, so the ability to dedicate the organization's resources to the organization is assured, even while the third-party provider handles other organizations in its overall cloud.

As with the other private cloud deployment models, some aspects are more difficult than others. On the less difficult end are security and resilience. Security is not a demanding challenge because even though the third-party provider is operating and maintaining the organization's cloud, it is the organization that has the ability to establish its security standards and its compliance protocols. Unlike the other two models, resilience is relatively easy due to the fact that third-party providers often specialize in the resources and operations necessary for cloud resilience. It's one of the reasons organizations take the dedicated private cloud route.

More difficult are all the other aspects of the cloud deployment model. Control, flexibility, elasticity, management, and mobility are all much more difficult and all because of the same reason – the fact that the third-party provider is operating and maintaining the cloud for the organization. In the dedicated private cloud model, the organization gives up a great deal of control and management capability because it is the third-party operator that is operating and maintaining the cloud. Flexibility is lessened to a great degree – potentially – because the organization must work within the parameters of the provider's architecture. Finally, the ability to scale up (elasticity) and the ability to be as mobile as needed are lessened because the organization is at the mercy of the provider's timeframe and ability to provide additional cloud capacity.[23]

Despite their costs, private clouds continue to be preferred by organizations for the security and control that are possible with private clouds. Additionally, organizations, as they grow, often find that private clouds are well within their budgets and so offer excellent opportunities for the organizations to control and to manage all aspects of their cloud.

The Hybrid Cloud

The third type of cloud is the hybrid cloud. It is designed as a combination of two or more of the three other forms of clouds (public and private, private and community, or community and public). In this structure, each of the two or more cloud systems remain as unique entities, but they are joined together by standardized or proprietary technology that makes possible data and application portability.[24] The hybrid cloud differs from the public cloud in that at least a portion of the configuration is internal to the organization. As such, it has both a private and a public component. The hybrid cloud also differs from the private cloud in that the hybrid cloud makes use of a private cloud for some functions and a public cloud for other functions. The hybrid cloud is very similar to a managed private cloud, but differs in that some operations are on the public cloud while others – generally those that are the most vital operating tasks[25] – are kept on the private cloud. "With shrinking IT budgets and rising business demands, companies are looking for an intermediate model of cloud computing technology. The hybrid cloud computing solution fits into this space."[26]

A hybrid cloud configuration can offer an organization certain features that will be attractive to those looking for the intermediate step between the public and private cloud. These features include:

• Scalability – hybrid clouds offer more scalability than private clouds because of the public/private nature (or public/community nature) of the cloud. As such, organizations can move non-essential operations to the public cloud portion which allows for maximum scalability while keeping the sensitive operations on the private cloud. By moving the non-essential operations to the public cloud, the organization is relieved of the storage requirements and thus can have a smaller, more cost-effective, but always scalable private cloud, with all the security elements inherent in the private cloud.

• Cost efficiencies – private clouds are expensive to build, operate, and maintain. Public clouds offer significant economies of scale and can offer low-cost, pay-as-you-go services. By bringing the two clouds together in a hybrid cloud model, the organization can use a smaller private cloud for the sensitive operations while using the much-lower-cost public cloud for the non-essential operations. For the organization, the cost savings can be very significant.

• Security – in the hybrid cloud, the security is placed where it is most needed – in those sensitive operations that are critical to the organization – through the use of the private cloud component, while those components needing less security are stored on the lower cost, third-party-maintained public cloud. Note that the public cloud does not mean no security, but it does mean less (see discussion of the public cloud above).

• Flexibility – hybrid clouds can be extremely flexible as they combine the best aspects of both the private cloud and the public cloud in terms of flexibility. They can deliver both the secure private cloud resources and the scalable, cost-effective public cloud resources that provide the organization with the ability to explore a variety of different operational avenues. The hybrid cloud also provides the flexibility to move resources quickly and easily from the private to the public cloud and back.[27]

The major concern of the hybrid cloud is, as to be expected, the five concerns discussed in the public cloud section of this chapter. Generally, for organizations considering a hybrid cloud model, security is the main concern. However, each of the concerns mentioned above in the public cloud section, as well as the security concern just mentioned, is ameliorated by the fact that the public cloud component in the hybrid cloud model is used for non-essential, non-sensitive operations of the organizations. For that reason, there is much less possibility of serious harm to the organization should something happen in the public cloud component.

The Community Cloud

These days, much less is discussed about the community cloud. Some writers include the community cloud as a separate cloud model.[28] Others consider it to be a subset of the private cloud model, replacing the organization with a community of some type.[29] Still others consider the community cloud to be a subset of the public cloud because of its potential use in government and industry.[30] Community clouds are designed for the exclusive use of a specific community of cloud participants or users that are part of organizations that have shared concerns. The community cloud can be owned, managed, and operated by one or more of the organizations in the community, but doesn't have to be. The community cloud can also be owned, managed, and operated by a third-party provider, or it can be a combination of one or more of the organizations along with the third-party provider.[31]

The important aspect of the community cloud is that the organizations forming the community (and it is almost always organizations, such as an industry, that uses a community cloud) need to have similar cloud requirements and have the goal of working together to achieve the objectives of the community.[32] Vertically integrated corporations, such as large holding companies (or television station group owners) could use the community cloud model to tie together the various portions of the organization to create a sharing capability across the various parts[33] (much like the links of a chain which was originally used to describe the radio networks in the 1920s).

The community cloud computing model, however, does not seem to be popular with CIOs of organizations. According to research from Cisco Corporation, only five percent of CIOs consider that the community cloud will be the dominant on-demand model by 2017.[34] Ian Cohen, group CIO at financial specialist JLT Group of London, says, "The problem with the fad for community clouds is that it is likely to be just that – a fad."[35] Nevertheless, given the right circumstances, the right groups of organizations, the right industries, etc., community clouds offer a viable alternative to the other three types of cloud computing models.

Summary

Whether there are four distinct types of cloud computing models or only three with the community cloud being a subset of either the public or the private cloud model (depending on who one talks with) is of little matter, compared to the enormous advantages the cloud models offer television in the 21st Century. Each of the cloud models offers opportunities for the television content providers, depending on whether it is an individual television station, a television station group owner, a cable channel, or a broadcast network. Chapter 7 will look at these opportunities in depth.

CHAPTER 3
CLOUD SERVICE MODELS

One of the reasons that the cloud has received so much attention is that it has changed how so many things work. No longer is software bought in a box off a store shelf at a local BestBuy, and most applications don't even need to be downloaded and installed anymore. Instead, the cloud has provided "Software as a Service" – the ability to run applications like word processors, email, finance applications, and more through web browsers.

Software as a Service, also referred to as "SaaS", is one of many "… as a Service" models that will be explored in this chapter. As will be seen, each of these models shares common themes:

- They don't require anything to be shipped, downloaded, or installed;

- They are accessible through the Internet and are used or managed through a web browser; and

- Rather than requiring that they be purchased in one lump sum, they can be paid for a little bit at a time – either for a fixed fee each month or every time the service is used.

The "… as a Service" family includes: "Software", as mentioned above; "Infrastructure", the computer hardware that is used to power the cloud; "Platform", which provides a collection of tools for creating a cloud service; "Data" which provides the information that powers applications; and "Television", which combines all of the above to create a new way of delivering and watching TV. The first cloud service model to be discussed is Software as a Service.

Software as a Service

In the early days of software, people would shop in stores or mail order catalogs to purchase software in a cardboard box that contained discs for installation. As the Internet came about in late 1990s, a new class of company arose called an Application Service Provider, or ASP for short. The typical ASP installed software from another company on a dedicated server as a way of outsourcing IT. All of the installation and maintenance of hardware and software was taken care of by the ASP.[1] The Software as a Service (SaaS) model grew out of ASPs with a few key differences: the company that develops the software is the one that makes it available for use online and the software doesn't require dedicated servers for each customer. More on the latter point about not requiring dedicated servers later.

There is a wide variety of companies that have SaaS offerings today, some of which are immediately recognizable consumer brands and others are behind-the-scenes players. Examples of SaaS companies include: Salesforce.com (CRM), Workday (Human Resources), Athenahealth (Healthcare), Dropbox (Document Management), Cisco WebEx (Desktop Sharing), Google Apps (Office Suite). The list is far too long to mention even all the major players; however, Bessemer Venture Partners offers a list of 300 top cloud computing companies, most of which are SaaS companies.[2]

It should be immediately recognizable that most SaaS products are as easy to use as creating a new email account on Google's Gmail or Yahoo Mail. They don't require a trip to the store to purchase, there are no discs or downloads for installation, they generally come preconfigured and are ready to use immediately, and they don't require periodic software updates.

Because the software lives on the SaaS provider's servers, there is a variety of new business models that are used for SaaS. Users are no longer required to spend hundreds or thousands of dollars (or more);

instead they can pay to use the software in small incremental pieces. In fact, because getting started with a service requires virtually no effort on the part of a SaaS provider, many providers start with a "Freemium" model – some limited version of the software is free and immediately available (frequently with advertising embedded into the software) – with one or more upgraded versions of the software available for purchase. Examples of freemium models include Dropbox (first 2GB free, 100GB for $9.99 per month[3]) or WebEx (3 people per meeting for free, up to 8 people per meeting for $24 per month, and other plans for more attendees[4]). The examples also show the "service" aspect of the Software as a Service business model. Software is made available for some low monthly, per-user, or per-use fee rather than offering a unlimited use license before the first use for a much higher fee.

Infrastructure as a Service

The second member of the "… as a Service" family is Infrastructure as a Service, or IaaS for short. IaaS refers to the servers that are used as part of the Internet, either to serve webpages, host databases, run firewalls, or provide other functionality. Prior to IaaS, companies would spend thousands of dollars on each server and install them either in their offices or in centralized networking facilities (referred to as "colocation centers" because servers from multiple companies were "colocated" there). Frequently, these servers would sit idle for large periods of time during off-peak hours, meaning that these companies had paid large amounts of money for servers that they weren't currently using.

The IaaS model enables companies to set up a new server nearly instantly. Just as Software as a Service is made available through the Web, so is IaaS. Users can sign up for an account and use a webpage to create a new "virtual server." To the end user, a "virtual server" looks the exact same as a server that they could have paid thousands of dollars for, but they don't have physical access to it. Every time a user starts a new virtual server, they don't know if it is the same physical hardware as the last time they started a virtual server or a completely different machine – but they typically don't care, as long as it gets the job done.

IaaS providers such as Amazon Web Services, Rackspace Cloud Servers, Joyent Compute Service, Microsoft Azure, IBM Softlayer/SmartCloud and others buy thousands of servers and put them in colocation centers with very high-speed networks and rent them out by the hour. Prices for running a server may range from two cents per hour to $12 per hour depending on how much computing power (CPU, memory, disk space, etc.) the server has. Likewise, other IaaS services have started to appear, the most common being Database as a Service which enables a new database to be created instantly and is typically charged by the amount of storage required for the database. More specialized tools with funny names like "Hadoop" (a "big data" processing technology, typically used for creating business intelligence out of huge amounts of information[5]) are also being made available by IaaS providers.

One important aspect of IaaS is "horizontal scaling." One of the interesting aspects of the Internet is that a person or company may never know when his or her funny cat video or the company's corporate website is suddenly going to be very popular. The last thing that a person or company wants is to be the most exciting thing on the Internet, but not have enough computing power to show everyone in the world what has been created. There are two ways that a person or company can ensure that (s)he/they always have enough computing power: buy bigger servers ("vertical scaling") that can handle lots of requests, or buy more servers ("horizontal scaling") and spread the requests out across a large number of similar servers. Because IaaS enables companies to start new servers nearly instantly, horizontal scaling means that they can always have the exact right number of servers running for the number of requests they are receiving.

IaaS does not necessarily have to be provided by third-parties, such as Amazon and Rackspace. As discussed in Chapter 2, the cloud can be any one of four different types: public, private, hybrid, or community. Large companies, such as Google, act as their own IaaS provider with large numbers of servers shared among the divisions of the company.

Platform as a Service

The Platform as a Service (PaaS) model is perhaps one of the harder models to explain because it is built for and used by the software engineers that develop Software as a Service. PaaS acts as a set of tools and building blocks for software engineers that are looking to build an SaaS product. PaaS is an extension of IaaS – it includes the servers that are required to run the software and PaaS providers manage the setup of hardware, storage, operating systems, software development tools, etc. To use an analogy, if SaaS was a car, PaaS would be the nuts, bolts, gaskets, doors, windows, and wheels that a mechanic would use to build the car. Continuing that analogy IaaS would be just the engine and transmission of a car and the mechanic would have to find and assemble all the rest of the pieces by himself.

The value offered by PaaS providers is that software developers don't need to hunt down every building block that they need to build their applications – it is all automatically managed by PaaS providers. That means that the major difference between PaaS providers is which pieces they have chosen to bring together. For example, Google App Engine uses Linux and open source technologies, as opposed to Microsoft Azure, which offers PaaS that is largely based on Microsoft Windows and other Microsoft-specific technologies.

The list of PaaS providers is long, and, in addition to Google App Engine and Microsoft Azure, companies include Salesforce.com (with their Force.com platform), Open Shift, Cloud Foundry, Engine Yard, and CloudBees. There is also some overlap with IaaS providers, such as the ones listed in the previous section on IaaS, that not only provide hardware but give users the option of deploying their design of PaaS as well. A comparison of PaaS companies can be found on paaslist.com.[6]

Unlike IaaS, which typically charges per minute for using hardware, the PaaS providers typically provide packages with some sort of monthly fee. Packages may include some pre-determined number of servers or computing power, a limited amount of storage, a limited number of users that can use the platform, etc. This packaging makes it easy to understand ongoing costs, where as IaaS costs (especially with horizontal scaling) may be hard to predict for any given month.

Data as a Service

The newly emerging Data as a Service (DaaS) model is driven by the spirit of a quote from business guru Tom Peters: "Companies that do not understand the overwhelming importance of managing data and information as tangible assets in the new economy will not survive."[7] Unless you are an engineer, it may not be obvious why data is important. Data is like a menu at a restaurant: a person doesn't show up to the restaurant to eat the menu, but the menu is critical in deciding what (s)he is going to eat. Every software application, including SaaS, depends on high-quality data to do its job – if the data in the application aren't good, then it doesn't matter how well the software is written.

Data as a Service companies provide services to manage data, ensuring that it is accurate, connecting data from different sources together, creating new data to create new value, and distributing data to customers so that they can realize the value of high-quality data. In the entertainment world companies like this have existed for decades, with companies like *TV Guide* and Tribune newspapers gathering information about television shows, verifying that it is accurate, and distributing it to viewers so they can make choices about which television shows to watch.

In the new "… as a Service" model, these DaaS companies make it much easier to get data than through newspapers or magazines. This information is available through Internet-connected services that enable DaaS providers, such as V2Solutions, to create and transform data so that companies and end users can get the most value out of it. DaaS providers may charge for their services by the amount of data that is created (the "volume-based model") or be based on the type and complexity of the data that is created (the "data-type model").

Television as a Service

Television, in one sense, has always been a service, so referring to Television as a Service (TVaaS) may seem redundant. However, what makes TVaaS new and interesting is the same thing that makes the other cloud service models interesting – its instant availability, being able to access the service through a web browser, and small up-front investments. For decades television has not been available without a set-top box that was installed in a home by the local cable provider or a DTH satellite television provider. This setup frequently required hours of labor, occasionally multiple trips, and hundreds of dollars of equipment[8] – not at all close to the instant availability and small up-front investments seen in the "… as a Service" models.

Over the last several years, streaming video has become commonplace with Netflix and YouTube accounting for more than 50% of Internet traffic.[9] To some extent, TVaaS is already available today – in much the same way that ASPs were the forerunner to SaaS. As industry moves towards the larger vision of TVaaS, content producers will be able to distribute content directly to the viewer without the need for intermediaries or distribution channels.

TVaaS isn't just about new technology or the "… as a Service" model applied to television – it meets a real consumer need. The very nature of broadcast television means that audiences are forced to watch what is most popular, not what would be the most entertaining to them. Marshall McLuhan's four immutable laws of media[10] have shown that given the opportunity, media will fragment until everyone has his or her own specific, targeted content. Because TVaaS enables more varieties of content to directly reach the audiences it was developed for, it has much higher value to end users. This is supported by the trends of decreasing production costs[11] and increasing discretionary free time for entertainment (driven by mobile devices). To realize this higher value for audiences, TVaaS combines all of the other "… as a Service" models mentioned above. Most important are DaaS and SaaS, which enable end users to find and watch television content.

Discovery of video is a significant challenge for TVaaS. Discovery means being able to browse or search through all available television shows so end users can find a show to watch. Cable television already has more than 190 channels[12] and it is frequently difficult to find the right show at the right time. YouTube has over 500 million channels[13] with 100 hours of new video being uploaded every minute,[14] making the problem of finding the right content to watch at the right time exceedingly difficult. Having the right data – the right menu for all the video that is available – is very important for TVaaS. The DaaS aspect of TVaaS ensures that all the videos have titles, descriptions, keywords, and other information that makes it easier to find the right video, similar to the "guide" feature of today's set-top boxes.

Similarly, SaaS is very important to TVaaS because it enables even the smallest content creators to become their own television channels. The beginning of this development can be seen in YouTube and their millions of channels, but SaaS doesn't require the use of a specific partner such as YouTube. In fact, three different models for the SaaS component of TVaaS can be seen today. YouTube is one model, which enables smaller content producers to upload content, create channels, and build an audience. The other two models for SaaS under TVaaS are TV-on-demand from traditional broadcasters and TV-on-demand from traditional distributors such as cable companies.

For the broadcaster's part, they frequently make content available through their websites or mobile applications for free. This can either be through a TV Everywhere type service (as described in Chapter 5) or as a free streaming video service. The important aspect of these services is that content from a single provider is made available through a single portal. An example of this would be HBO Go, which makes content available to their customers for free (provided that their customers login using a username and a password from their local cable companies). Another example would be abc.go.com, which provides free streaming of ABC Network's television shows as a free, turn-key service.

The traditional distributors have a similar SaaS play for TVaaS. An example of this would be Comcast's Xfinity On Demand, which allows users to watch a collection of videos as part of their Comcast subscription. An important aspect of the Xfinity On Demand service is that Comcast is aggregating a collection of videos across different content providers, simplifying the discovery experience for an end user.

The future of television may see a blend of these SaaS-for-TV models. It may very well be that content providers large and small make their content available through web services, and an aggregator would collect all these different services to aid users with discovery. Note that an aggregator is not the same as a distributor, because the video would still be delivered directly from the content providers – it is just the discovery of content that is provided as SaaS. Real aggregators may come about through the mechanisms described in Chapter 5, where web services can be mixed and matched by nearly anyone to create new TVaaS offerings.

Summary

This chapter has explored five different cloud service models: Software, Infrastructure, Platform, Data, and Television as a Service. Each of these share the common attributes of the cloud that they don't require installation, can be used or managed through the web, and are available as an on-demand service with monthly or per-use billing. The combination of all of these services culminates in a new cloud model for television – Television as a Service. This new TVaaS model enables instant access to more content than ever before through the web and has the ability to change how the entertainment industry and audiences think about television.

CHAPTER 4
ADVANTAGES OF THE CLOUD

Whether an organization chooses to use the public cloud model, the private cloud model, the hybrid cloud model, or the community cloud model, there are several distinct advantages in using the cloud over various other means of storage of data and the variety of uses those data require. While the advantages were mentioned in the discussion of the different cloud types in Chapter 2, this chapter takes a look at the major advantages inherent in all cloud models to one degree or another. The advantages that will be discussed are security, flexibility, mobility, portability, cost, scalability, and – one not mentioned earlier – the environmental or "green" advantage of the cloud.

Security

Security is one of the major concerns of organizations who are using or considering using the cloud for any purpose. Determining the type of cloud model to be chosen determines the amount of concern there is for security. For the public cloud model, security is a major concern for an organization because the security requirements are left to a third-party provider. At the other end of the cloud model spectrum, the self-run private cloud model causes the least concern for organizations because they are running the cloud completely and so have complete control over the security layers built into the cloud. Between those two end-of-the-spectrum models, the other types of private cloud, the hybrid cloud, and the community cloud have varying degrees of concern over security with the order of magnitude of concern going from less to more in the order of the cloud types listed above.

However, regardless of the security concerns associated with the various cloud models, the cloud offers superior security to other means of storing information. For instance, storing data on a local computer or on local servers (for an organization) subjects that data to being lost due to a hard drive or server crash. Hard drives do crash and crash often enough to be of significant concern; local servers crash as well. Whole industries have been built around the protection of data, from portable hard drives that fit in a person's coat pocket or easily on a desk and hold terabytes of data, to small flash drives used not only for storage backup, but also for the ease of portability that they offer.

While it is unlikely that an organization would back up its entire data set on a flash drive, small companies do use the portable hard drive as their security from hard drive crashes. In addition, there are other alternative forms of storage that can be used, including CDs and other such devices that offer low-cost, but limited protection to users as a backup for their data on the local computer or local server. However, each of these various devices has its problems as well. External hard drives can crash just as a computer's internal drive can. They can be lost or, more likely, stolen. Flash drives can get broken, lost, or stolen even more easily than external drives because they are not built for rugged use, and CDs can get scratched beyond repair. When storing information on these alternative devices, it is generally a good idea to back up the information on several of the devices in case something goes wrong with one or more of them. At that point, the costs begin to rise and a good deal of complexity is added in remembering where the different devices are stored, because an organization small enough to use such devices to store its data would need to have those devices in different locations with different amounts of security added to each of them. Further, backing up data then becomes more complex because any changes in data would have to be changed manually – most likely – for each device, a time-consuming and, sooner or later, exceedingly frustrating operation.

When an organization's information is stored on a cloud, those data are not subject to the types of damages that can occur with external hard drives, flash drives, CDs, and the like. It can be a real annoyance and an additional cost should a notebook computer get dropped or a car run over a flash drive, CD, or portable hard drive. However, when a business' work is stored on the cloud, it can be easily retrieved from anywhere, unlike from a destroyed computer or portable storage devices.

Further, clouds are safer because – especially today – companies offering cloud services design their clouds with layer-upon-layer of security protection, in addition to the redundancies mentioned earlier. A white paper for policymakers by the Software & Information Industry Association reports, "While there is some fear of greater risks with cloud computing because of the cultural change in relinquishing direct control of the IT infrastructure, there is a much less recognized reality that cloud computing, by nature, provides for an environment inherently superior for applying critical security measures."[1] The white paper goes on to enumerate what it refers to as "key security practices" that clouds can improve upon. These practices are

- Detection – cloud computing is about the linking of millions of security nodes together. The linkages make it possible to better detect cyberthreats of all kinds.

- Remediation – cloud computing makes it possible for cloud service providers, working with security providers, to implement solutions to defeat malware much quicker than having to load the "fix" onto multiple machines or devices. Remediation is a critical component of cybersecurity and a cloud allows for faster response times.

- Prediction – cloud computing makes it possible to build into the system predictive abilities to identify and block those who make it a point to develop and disseminate malware. If it's possible to predict the likelihood of who is attacking the system, it is much more likely to block and defeat the attacker.

- Protection against end user breach or corruption – cloud computing makes it possible to eliminate or minimize end user problems, most often either data breaches or data corruption that results from lost or stolen portable computers or other mobile devices or drives. These threats can be minimized or eliminated because the data are centrally stored and the network is continuously and automatically analyzed and protected.[2]

While it is always possible for a system to be "hacked" no matter how sophisticated its computer architecture is, cloud-based storage provides the best opportunities for protection of vital content of all types. However, it must always be remembered that "security is a collaborative effort between the vendor and the client."[3]

Flexibility

Flexibility appears to be one of the more difficult advantages to define. For some, the term is equivalent to "scalability," while others consider flexibility to refer to "mobility." Still others consider all three terms to be separate and distinct. Despite some differences in the defining of the term, for purposes of this book, the three terms will remain separate advantages of cloud computing models. Flexibility here means the ability to be agile,[4] to explore different operational avenues,[5] and to run multiple business processes that are different in nature simultaneously.[6] It allows the company to quickly and easily re-orient resources during peak traffic periods, moving less important processes or data sets to one portion of the cloud to make room for newer, more important processes or data sets. Flexibility also makes it possible to move processes and data sets from one cloud to another. The ability to move and manage all the various day-to-day operations, including those critical and those less important to the immediate success of the organization, is central to the concept of the cloud computing model.

The advantage of flexibility is also the advantage of speed. The flexibility that a cloud environment brings to an organization makes it possible not only to move the various aspects of the organization's operations but to do so much faster than storing and retrieving those aspects when contained in any other form or on any other type of device. Furthermore, consistency in terms of runtime environment across multiple clouds within the organization's structure (private to public, public to private, etc.) guarantees that the organization does not need to retest and requalify anytime a redeployment is called for.[7]

Mobility

For most organizations these days, mobility is one of the major keys to the success of the organization. Employees travel. New offices are established in a variety of locations that must be tied together to work seamlessly for the success of the organization. International locations are often established. These, too, must work seamlessly with all other locations and the home office for the organization to succeed. The mobility of the cloud computing model makes the mobile business environment possible in a way that is superior to any other. As long as a person, office, etc. has a computer or some other device capable of connecting to the Internet, and Internet connection capability, that person, office, etc., will be able to access the organization's cloud to retrieve whatever file or information is needed quickly and easily. As long as there is Internet access, businesspersons, offices, etc., can access information from home, on the road, from client's offices, during meetings, or whenever the need for that information is required. Further, with the ability to connect to the Internet from anywhere and to reach the organization's cloud architecture, businesspersons today are not limited to their computers, but can access and conduct business from their smartphones or, much more likely, their tablet computers.[8]

In addition, by using one of the various cloud deployment models, there is the added flexibility of collaboration. The organization of today is highly mobile, combining employees into teams and workgroups from offices and locations around the world. The cloud makes it possible for those team or workgroup members to collaborate with each other on a particular project, working together on files and documents even when physically present in locations scattered across the globe. The team or workgroup members collaborating on a particular project can view, revise, and edit each of the team or workgroup member's work, simultaneously or not, from their various locations, both in the U.S. and abroad.[9] The mobility advantage of the cloud is crucial both today - and even more so in the future – for both transnational companies as well as small and medium sized enterprises (SMEs) with partners in other countries. The mobility advantage of the cloud cannot be overestimated, and its importance will only be enhanced in the future.

Portability

In the early days of the development of the World Wide Web, well before the notion of "cloud computing" existed, the idea of portability was already considered to be an important advantage to organizations. Even at that early time, the Web was being thought of as a place for storing excess or archival data. The use of the Web for storage made portability an advantage because the thinking was that while a person or an organization could store content on the Web, the usefulness of the content would only be available when it was downloaded to a local device, whether a computer, a CD, a flash drive, external hard drive, etc. The downloaded material would then be carried on the alternative device wherever the businessperson was going, then used in the manner planned. The idea of accessing and using that content directly from the Web, or – in today's term – the cloud, developed as a later consideration.

Today, of course, the possibility of downloading content from the cloud still exists for businesspersons and organizations. However, there is, in reality, only one real scenario where the need for downloading content becomes necessary. That scenario is when no Internet access from broadband, Wi-Fi, or cellular will be available at the location where the downloaded content will be used. As long as Internet access is available through some means, the need for downloading content does not exist. Further, downloading the content to a local device brings into play all the safety concerns mentioned in the first advantage earlier in this section.

Cost

One of the main benefits of the cloud is the reduction in costs that can accrue to an organization. Whether operating one of the three private cloud models, a hybrid cloud model, or making use of a third-party public cloud service, most likely – but not always – an organization will realize significant cost savings. In the article "5 Financial Benefits of Moving to the Cloud," author Richi Jennings lists the following as benefits:

• Fully utilized hardware – Cloud computing brings about natural economies of scale whether for small and medium enterprises or large transnational organizations. Especially when utilizing a public cloud, the "pay-as-you-go" method of financing makes the cost savings over operating a large data center extremely significant, even for the smallest of businesses. Combine the cost savings with the ease of scalability, and the costs savings increase further. The organization is fully utilized in its hardware needs and can smooth out its peak and valley demand times.

• Lower power costs – Costs for the energy to run the cloud computing services, because they are shared among a number of users, is lower for the individual organization than if the organization were to operate its own data center. The obvious exception to this cost savings is the "self-run private cloud." However, even with the self-run private cloud, the energy costs could also be reduced if the organization fully utilizes its hardware. For most organizations, using one of the other cloud models, the energy costs of the cloud model can be substantial.

• Lower people costs – Budget expenditures for personnel to operate and maintain data centers can be extremely expensive. Depending on the cloud model the organization chooses, those personnel costs can be reduced in increasingly larger measures. The more internal cloud resources that are needed, the higher the costs, but every cloud model except possibly the self-run private cloud model benefits in the area of lower personnel costs because at least a portion of the organization is operating on a third-party provider's cloud.

• Zero (or at least lower) capital costs – Depending on the type of cloud model an organization adopts, the costs can be minimal to substantial. Private clouds involve extensive capital expenses (CAPEX) for the build-out and start-up of the cloud service, and then operating expenses (OPEX) to operate and maintain the cloud service. However, for organizations choosing the public cloud option, especially those choosing a pay-as-you-go option, CAPEX costs are zero and OPEX costs are determined by the amount of cloud space needed and for the length of time that space is needed. For hybrid and community clouds, CAPEX costs do exist for either the private portion of the hybrid cloud, or the organization's part of the community cloud build-out and start-up, but those costs can be, and often are, significantly less than the CAPEX costs of a private cloud.

• Resilience without redundancy – This benefit is specific to all clouds but the private cloud model. Operating a system of internal servers, an organization must build its redundancies through numerous duplicative pieces of hardware, much of it lying idle in case there is a need to replace a defective server or server rack. The extra hardware, plus the necessity of building redundancies into the system through the use of additional backup server racks is expensive to build and to operate. Moving a portion or all of the organization's operations to a third-party cloud service removes the need for on-site redundancies. The move to the cloud service allows for potentially numerous backups of data, providing the resilience needed without the CAPX and OPEX required for on-site servers.[10]

A recent study of 1,300 companies in the U.K. and in the U.S. shows the important economic impact that cloud computing can have on costs and cost savings. The results showed that 88 percent of those companies using cloud services found they had cost savings and 56 percent reported increased profits. Additional results found that 60 percent of the companies were able to reduce IT resources, while 62 percent of those companies reporting cost savings reported reinvesting the savings back into the business to expand and innovate.[11] At the same time, not all organizations – especially small-to-medium-sized ones – may save money in the long run.[12]

Scalability

Scalability is also a key benefit of using one of the cloud computing models. Depending on the cloud computing model chosen, scaling up or down can be extremely easy and quick, making the cloud both cost-effective and agile. This is especially true for those organizations that make use of third-party cloud providers in any of the cloud computing models except for the self-run private cloud model. A look back at Chapter 2 shows just how simple it is for the various models utilizing a third-party cloud service provider to increase or decrease the amount of server usage the organization needs and the ease of making those changes.[13] The ease of scalability also allows an organization to maximize its agility without expensive changes to its existing IT systems, and protects against system overload.[14]

Green Computing

The use of the cloud is being touted as an environmentally friendly option as it can reduce an organization's carbon emissions significantly.[15] Notably today, they are designed both to reduce the consumption of electricity as well as reducing the emissions that can damage the environment.[16] Many international companies are choosing cloud options today because of the green credentials cloud computing makes available.[17]

There are several environmental benefits of cloud computing. First, the third-party cloud providers, by investing in large scale data centers, allow each client organization to reduce its own data center waste.[18] Second, material waste is reduced in relation to the reduction of hardware needed by each client organization.[19] Third, to go along with the first two, because client organizations only use the server space they need, they are able to reduce their carbon footprints.[20] By using cloud computing, organizations generally can reduce their energy consumption and carbon emissions by at least 30% over on-site servers, with small organizations getting the greatest benefit through cutting their energy use and carbon emissions by as much as 90%.[21]

Summary

Cloud technology, clearly, is in a state of rapid deployment and improvement – it is a technology for both today and for the future. Cloud technology provides organizations with a number of important benefits – benefits that can be applied to 21st Century Television. The newer media organizations – both hardware and software companies – are already enjoying the benefits of their deployment of cloud computing technology to their television and video offerings. The continuing deployment and ever-increasing improvement of cloud computing technology provides the newer media organizations (Netflix, YouTube, Amazon, etc.) with ever increasing opportunities for success. The more rapid the deployment of Television In The Cloud by the legacy media, the greater is the likelihood for growth and innovation of those organizations in this revolutionary 21st Century Television universe.

PART II
TELEVISION IN THE CLOUD

CHAPTER 5
BUILDING BLOCKS OF TELEVISION IN THE CLOUD

Before discussing the building blocks of cloud television, it is important to understand that one of the important aspects of the cloud is what is called a "Service Oriented Architecture" or SOA. The term is technical, but the concept is not difficult to understand – it is all about interchangeable parts. To use an analogy, prior to the cloud most technology environments were like a fast food restaurant: they specialized in doing one thing and doing it well. If a person wanted hamburgers (s)he was in luck, but don't dare to dream of tacos or gyros. At its core, the cloud is more like a farmers' market than a fast food restaurant. There is a wide collection of vendors providing services – fruit, vegetables, meat, cheese, nuts, honey, and so on. If suddenly customers need vegetarian food, the change is easy – get rid of the meat vendors and bring in more vegetable vendors. Likewise, if customers want barbecue, similar adjustments can be made.

Service Oriented Architecture means that interchangeable services can be swapped in or out to remain flexible in how one gets to the end goal, and even flexible about what the end goal is. This chapter will explore a Service Oriented Architecture as applied to television, starting with the most modern and common language for the cloud: web services.

Web Services

While purists would argue that web services are one of many languages for the cloud,[1] the reality is that almost every cloud model today uses some aspect of web services. These web services are similar to the technology that feeds today's browsers. The only difference is that web services aren't something that an end user would typically see. Browsers are based on HTML, the Hyper Text Markup Language that is not that different from word processors that make it possible to create bold text, insert images, and change the style of the fonts. The purpose of HTML is to create documents for people to read.

Web services are based on a similar technology called XML – the eXtensible Markup Language. Rather than focusing on creating documents for people to read, XML can be used for nearly anything. In the case of web services, XML is used for sharing information between services. If an end user wants to know what television shows are playing tonight, (s)he might visit http://tvguide.com, but that website is written in HTML so that it is easily read by people. If a server wants to know what is playing tonight, it would visit http://api.rovicorp.com/data/v1.1/video/schedule, and receive roughly the same information in XML format – the language that is much easier for computers to use.[2]

Web Services – Building Blocks for the Cloud

Television in the Cloud is nothing more than a combination of these web services – the building blocks of Cloud Television. There are web systems for all the different pieces required to create Television In The Cloud, from production systems and other enterprise systems (finance, legal, CRM, etc.) to all the rest of the television specific systems. By mixing and matching the different web services for billing, account management, content libraries, streaming video, advertising insertion, linear playback, and others, any content distributor can create a TVaaS offering. Part of the appeal of using web systems for creating TVaaS is that the different web services can be used or not used depending on the business model or technical approach that any TVaaS provider chooses. For example, if television shows are made available for free, there is no reason to integrate with web services for authentication and authorization. Likewise, if a service is on-demand, there is no reason to integrate with a web service for linear playback. As time goes on, it is likely that there will be more web services than are available at the time of the writing of this book, giving TVaaS providers even more flexibility in how they mix and match web services to create their products.

The rest of this chapter takes a look at the web services that are the building blocks of Television in the Cloud.

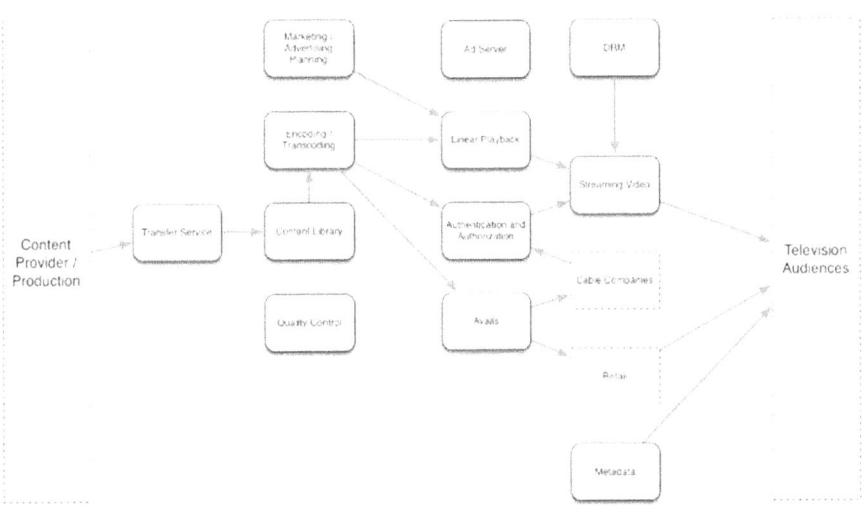

FIGURE 1 – The major systems and services of a broadcast network

Content Libraries

The traditional content library for broadcast networks is the archive. An archive immediately paints a picture of dusty vaults of antiquities that have long since been forgotten. For television this is true to some extent, because all the television shows ever aired will be stored in the archives – including old cans of celluloid film and tape cassettes from the more recent past. However, archives are also the place where television shows are stored before they are aired. When content is received from a production company, it is typically stored in the archives and transferred from the archives to the various systems that will be airing it. (This may be true to a varying degree for different broadcast networks – some broadcast centers will work directly with the production companies or post-production houses to get access to content in parallel with its delivery to the archives.) The archives are typically a company-wide storage facility for what is referred to as "masters" – the final television shows. There may also be "sub-masters", which are slight changes to the masters that have long-term value and that need to be saved. Examples of sub-masters would be a show that has been dubbed or subtitled for German or French.

A similar system is used outside of traditional broadcasters, where online video distributors and TVaaS providers use content libraries. These libraries may be as simple as a shared drive connected to a network, with a different folder for each television show. There are also more sophisticated systems called Digital Asset Management (DAM) or Media Asset Management (MAM) systems, although the difference between a DAM and a MAM is sometimes hard to distinguish. DAM software, such as Evertz Mediator, 5th Kind, the REACH Engine from Levels Beyond, or OpenText Media Management, offers its own set of web services and varying degrees of customization, all to make storing and distributing content as easy and efficient as possible.

The common attribute of content libraries for TVaaS is that they all include web services for submitting, searching, and retrieving television shows. These web services may be used by applications that enable employees to upload, download, or search for videos, or they may be used by other services, such as transfer services, transcoding services, and linear playback systems – all of which are described below.

Transfer Service

Professional grade video files are large, ranging from tens to hundreds of gigabytes.[3] (For reference, at the time this book was published, the largest iPhone or iPad models are 64 gigabytes. Depending on the television program, those models might not be able to store even one professional video.) Even with the highest-speed networks available they may take minutes or hours to transfer.

Transfer services are responsible for moving around large files between outside vendors, such as content providers, post-production houses, content libraries, and internal divisions, such as marketing and distribution operations. The transfer services will ensure that the files are delivered, no matter what circumstances occur, and report back both to the sender and to the recipient with status updates and error messages so that the transfer can be monitored.

The transfer service will offer web services for all other services to use and are frequently integrated with almost every other building block in a TVaaS system. One of the most important places where the transfer service is used is to send and receive files from the archive, and it is especially important for sending files from the archive to the distribution centers, such as the broadcast operations center.

Encode / Transcode Service

Whether it's a television, a mobile phone, a tablet, or a desktop computer, nearly every device has its own preferred format for a video. Each television show may require up to 30 different formats for all the different places it will be viewed. The purpose of a transcode service is to convert the master television show in the archive to the different formats that different screens require.

Transcoding very large video files is a very expensive service – requiring either large collections of servers or smaller collections of specialized hardware. Because nearly every division in a broadcast network requires transcoding for their own internal or external needs, transcoding is increasingly becoming a centralized service that is shared across divisions. Traditional "file based workflows" (described in Chapter 6) have enabled integration with transcoding services in the past, but these services now offer web services and can be (and are being) integrated with nearly every other system that handles content across a TVaaS system.

Linear Playback

Linear playback is one of the central aspects of traditional television service. For decades, television programs have been broadcast back-to-back based on a pre-determined schedule, often creating television channels that offer television programming 24 hours a day.

The traditional systems that support linear playback are trafficking and automation systems. They are the systems run by the broadcast operations departments that receive the television programs and commercials (trafficking) and play them in the right order (automation). The traditional vendors for these systems have moved from tape-based playback to using digital files, and are increasingly offering web services for integrating with other systems at the broadcast network.

Traditional broadcast products, such as Harris D-Series, Evertz Playtime, and Sintech OnAir, are increasingly moving away from tightly integrated products, to products that offer web services that seamlessly integrate with products from other companies. In the newer TVaaS systems, linear playback may be provided by video platforms such as Ooyala or Ustream. Both classes of linear playback systems offer a combination of SaaS and PaaS building blocks. The SaaS part of linear playback systems enables TVaaS providers to create schedules and determine which video file will be played when. The PaaS part offers web services to integrate the linear playback system with content libraries, transfer services, ad sales systems, and more.

In addition to the web services consumed by linear playback systems, they also offer a number of their own web services. Examples of these web services include the status of playing back a video, lists of which shows have been played and when they have been played, which advertisements have been played, and so forth. Given the central role of trafficking and automation, the number and type of these web services is likely to grow over time.

Authentication and Authorization Services

The Authentication and Authorization services, or A&A for short, are relatively new to broadcast television. They have come about through subscription services, such as those offered by a cable company, to enable end users to access broadcaster or cable network content outside of the typical cable television environment. In fact, CableLabs, a non-profit organization funded by the cable companies to help them grow their industry, developed the detailed technical specifications for Authentication and Authorization. (The specifications have since been taken over by the non-profit Open Authentication Technology Committee (OATC), which is responsible for seeing them through to implementation and adoption.)

The way authentication and authorization typically works is that a user will visit a broadcast website or use a broadcast or cable network's mobile application and choose a show to watch. The user will then be prompted to enter his or her username or email address from the cable provider and his or her password on the cable system. The broadcaster or cable network will use this information to communicate through the web services with the cable company to verify that the user has the correct user name and password (authentication). If it turns out that the user has a valid account, the broadcaster or cable network will use a second set of web services to verify that the content (s)he is trying to access is part of the service (authorization).

The number of applications and websites that implement these new web services is growing, including support from the FoxNow and HBO-Go apps (among others). These services are generally referred to as "TV Everywhere" and are intended to enable consumers to watch television even while they are away from their television sets.

Avails Service

The purpose of the avails service is to let customers of the broadcast network know that content is available to their customers that have signed licensing deals. These customers may include companies such as Netflix, Amazon, Hulu, CinemaNow or Xbox (among others) – generally the online retail stores that rent or offer subscriptions to television shows. This entire class of stores is generally fairly new and the web services for supporting them are just now under development. Traditionally, companies that have licensed content will receive a spreadsheet or an email letting them know what content is available to them or that new content is available. These notices are referred to as "avails" for short.

Only recently has the non-profit Entertainment Merchants Association (EMA) begun development of specifications that will lead to web services for avails notifications.

Metadata and Discovery Services

Before television viewers can watch videos, they first must decide which videos they want to watch. In parallel with the emerging avails services, there are also new standards emerging for studios to distribute metadata to their partners. This metadata includes information such as a synopsis of the show, a list of cast and crew, genre for the show, artwork for the series and the episode, and generally any other information that would be required to help convince a consumer to buy a show. These metadata services take the place of DVD boxes in traditional distribution – users used to be able to browse through aisles of DVD boxes at their local retailer to decide which shows to buy. Now, they browse through metadata in online stores.

The metadata distributed to retailers is fairly static. For their own websites and mobile applications, broadcasters will have their own metadata services that go one step further by offering the ability to discover content through search, browsing, and recommendations. Content providers hope it is intuitive what these search web services do. Because they are similar to Google, users can enter a search term and receive a list of videos that matches their requests. Likewise, browsing through a list of shows should be as familiar as reading a *TV Guide* or a newspaper listing of shows, and web services exist to provide this same information one page at a time.

Recommendations are the newest and perhaps the most interesting of the web services. These web services will use information about a user's viewing habits (and sometimes the viewing patterns of other users with similar tastes) to suggest shows for a viewer to watch. For anyone that has ever used the Netflix website, this category of discovery should be recognizable. Netflix puts recommendations for dozens of shows front and center in their service rather than requiring users to navigate through long lists of shows to find something to watch.

These discovery services are typically created for the private use of the broadcaster; however, retailers and cable companies will usually have their own web services for discovery. After a user discovers a video (s)he wants to watch, the viewer can usually start streaming it right away.

Streaming Video Services

Since the late 1940s, television has been broadcast over the airwaves to audiences worldwide. In the United States, over-the-air broadcasting has largely given way to cable companies that offer more reliable service delivering programming through a wire (cable) for a monthly fee. Both the over-the-air broadcast and subscription cable services are slowly being replaced by similar services that use streaming video over the Internet for distribution.

Streaming video systems typically have two components: a server that integrates with a content library, and a video player that receives a stream of video over web services that are provided by the server. The video player may be found on websites, mobile applications, gaming consoles, set top boxes, or even integrated into connected television sets.

The web services that the server makes available to the client typically include functionality to start playing a video, stop, pause, fast forward, rewind, jump to a specific location, etc. There are numerous streaming video solutions, including Adobe Dynamic Streaming for Flash, Apple HTTP Adaptive Streaming, Microsoft Smooth Streaming, Verizon upLynk, and even open standards such as HTML5 and MPEG-DASH.

One other aspect of the streaming systems is their ability to detect network conditions and adjust the quality of the video to use more or less bandwidth. This "adaptive bitrate streaming" technology ensures that a video doesn't stop playing should bandwidth reduce, and it relies on a number of web services to report network conditions to make the appropriate adjustments.

Ad Server

In order to monetize Television in the Cloud, an advertising server (or ad server, for short) may be required. The ad server may integrate with a content library, or have its own content library of commercials that will need to be scheduled into a show. In linear playback systems, the advertisements may be pre-scheduled for when each commercial should be seen – before the show, after the show, during the first commercial break, etc. In on-demand systems and more modern streaming video systems, it may be pre-determined when the advertisements will air (for example, at 5 minutes and 14 seconds into the show), but which commercials and the number of commercials to air can be decided as the show is in progress.

The ad server plays a large role in determining which commercials will air. It will receive information from the video player through a number of web services about who is watching the show – their age, where they are currently, what device they are using, etc. The server can use this information to send targeted advertising to each user – advertising that has been specifically selected for that user that has the highest chance of being most relevant to him or her. The ad server may also interface through web services with billing systems, to keep track of which commercials aired. This information is ultimately used to charge advertisers whose commercials actually aired.

Digital Rights Management (DRM) Services

In order to prevent piracy, a system is usually set up to ensure that only end users that have the rights to access content can view it. These Digital Rights Management (DRM) systems are typically included as part of streaming video systems. For example, Microsoft Silverlight includes Microsoft PlayReady DRM and Adobe Flash includes Adobe FlashAccess DRM. These DRM systems may also be used independent of streaming video for video downloads.

Each DRM system typically has a set of web services for verifying a user and his or her permissions to access a piece of content – similar to the Authentication and Authorization services mentioned earlier, but without the need to communicate with cable systems to get information about subscriptions. DRM systems are also increasingly being integrated with Ultraviolet, an industry-wide service for managing a user's rights across all studios. Part of Ultraviolet is the "coordinator" service, which has a set of web services for submitting rights to be remembered (for instance, when a user purchases a digital video) and for retrieving rights (for example, when a user goes to watch a video that they have previously purchased).

Quality Control Service

After spending millions of dollars to produce a show, ensuring that the quality of the content isn't eroded by technical glitches is very important. To that end, there are entire departments dedicated to quality control and checking for defects in shows. These departments are increasingly moving to automated QC software such as Interra System's Baton that will check for flash frames, video dropout, pixelation, or other common errors. These systems can be accessed through web services to submit files for quality checking.

Part of the value of having a web service for quality control is the ability to have quality control as a shared resource. Currently, files will be put through the same QC checks multiple times as they pass through different organizations, despite the fact that they are all verifying the same file. By centralizing quality control, duplicate quality control efforts can be eliminated.

Operational Support Services

Outside of the systems necessary to make content available to end users, there are a number of management systems that consume web services in an attempt to monitor and control the operational aspects of running a broadcast network. These include analytics and business intelligence, contract compliance and digital store check solutions, network operations centers and technical support, and a number of others.

Summary

As has been discussed in this chapter, there are a large number of building blocks for a cloud-based television network. Some of the major systems and services include transfer services, archives, transcoding, authentication and authorization, linear playback, avails, metadata and discovery, streaming video, DRM, quality control, and operational support. Each of these building blocks both offers their own web services, and, at the same time, consumes the web services of other building blocks to create a flexible service-oriented architecture.

CHAPTER 6
MOVING TELEVISION TO THE CLOUD

In the future, television is likely to be very different from how it is today. Since the beginning of television, the underlying broadcast technology has required audiences to watch the same show at exactly the same time. Watching television shows on a specific schedule one after another, the so-called linear programming model,[1] has become engrained in the television viewing culture.[2] Only in the last decade have DVRs freed the viewer from watching television on a specific schedule, although the shows are still delivered in the same linear fashion. This chapter looks at the fundamental changes that television is going through right now and how broadcast networks are in the process of moving television to the cloud.

Business Models: Past and Present

In order to understand the changes that are underway, it is important to first look at where television has been. The original television stations in the late 1940s – NBC, ABC, CBS and, much later, Fox – were all the video equivalent of radio stations.[3] They were free to watch and the broadcasters made their money from advertisers that were looking to reach mass markets. By the early 1950s, communities had realized that they could share an antenna on top of a hill and run the signal to their houses down below for better reception, which became the beginning of the cable television model where consumers would pay for access to cable television as well as still viewing the broadcaster's commercials. These are the first two business models for television: ad-supported and subscription. The third business model – transactional – came about with the advent of VHS and the ability to rent or buy movies and television shows. Television is now offered under all three business models, although ad-supported broadcasts and paid subscriptions are probably the most commonly associated with television. Part of the reason for television being less associated with transactional business models is the windowing systems that studios have used for years.

Content Windows

As television shows and movies get older they are less valuable – a movie that someone might pay $20 to view in a movie theater today might only be worth $5 to rent next year or maybe the viewer might just wait to watch it on television for free the year after that. The studios have long used this system of selecting which distribution channel and which business model should be used for their shows to maximize their profits.[4]

In the world of broadcast television the first airing is the premium window – because there have traditionally been large audiences that will tune in for the latest episode in their favorite series and sit through commercials that advertisers paid significant amounts of money to put in front of those audiences. The second window has traditionally been purchasing content, where VHS tapes or DVDs are available through retailers for a price north of $10 or more, finally followed by content being made available for rental for a few dollars. The most popular of shows would also enter syndication, where other broadcast networks could pick up previous seasons of a show as reruns that would be broadcast to their audiences.

In the past, these windows have been months apart with television shows typically showing up for purchase after the end of a season. With the increase of video viewing on the Internet, the time between these windows has now shrunk to hours. Within 24 hours of the first airing of an episode, broadcasters will typically make an episode available for purchase and/or rental in parallel with its first international viewing. There has also been another window that has emerged – subscription video-on-demand (also referred to as sVOD) – which is when Netflix and similar subscription services make content available for on-demand viewing. The sVOD window is typically at the end of the season.

The emergence of Television In The Cloud with these compressed windows, along with the new "digital supply chain" for distributing content through the Internet, has required broadcasters to move faster, be more efficient, and re-invent the ways in which they create and distribute content.

The Broadcaster's Private Cloud

As explored in Chapter 5, the broadcast networks are increasingly moving towards Service Oriented Architectures where a number of building blocks are tied together through web services. Despite the fact that the traditional broadcast systems, such as trafficking and automation, are increasingly using web services, they are still the same at their core as they have been for decades. The more fundamental change for broadcast networks is the digital distribution channels that they use for retail, rentals and subscription video-on-demand.

For the last decade, broadcasters have worked on implementing a "file based workflow" – using computer files for editing and sharing videos rather than shipping videotapes to and from stations. This has been a huge operational benefit for broadcasters that can now move content faster with less overhead. The next step in this evolution is the "digital supply chain", which moves around not only digital files, but uses digital systems to communicate about schedules, tasks, timelines, and all the operational aspects of a broadcast business. Groups such as the Hollywood IT Society (HITS) are bringing together studios and technology vendors to try to develop common solutions for sharing information and moving content through the cloud.

The creation of a "digital supply chain" between a broadcast network and its vendors enables broadcasters to move toward the vision of Television In The Cloud. By automating the supply chains, broadcasters are increasing the amount of information available to consumers and reducing the number of intermediaries between content production and audiences. This means that audiences are better informed, find more content, have new interactive features and supplemental content available to them, and have increasingly more influence on the types of shows that are produced.

Video Management Platforms

Streaming video starts with an online video management platform, such as thePlatform, Ooyala, Brightcove or Kaltura. These platforms not only store the video files before they are streamed over the Internet, but they offer other functionality to assist in the management of the video assets. Some features commonly found in these platforms are windowing terms for on-demand content (i.e. - when the videos should be available publicly), metadata management for integrating the right metadata with the right video, search functionality, analytics and dashboards for operational reporting, and much more. These video platforms will also typically serve as the backbone for the rest of the video distribution process, tying together ingest and transcoding, streaming video servers, account management, digital rights management, advertising insertion, and other functionality that is key to delivering the content to the end user.

Billing and Account Management

For transactional or subscription services, a user has to pay to access the content. Billing and account management systems keep track of usernames, passwords, subscription information (such as whether a user has subscribed to HBO as part of a package or not), and generally confirm whether or not a user is allowed to view a piece of content. Note that for advertising supported content, this isn't always an issue, because the content may be free and available to everyone without any billing or any need to login to an account. After a user has sent his or her user name and password to the system to confirm that (s)he has the right to watch a video, the viewer can begin streaming the video.

Ingesting, Transcoding and Chunking

Before a video can be streamed to an end user, it has to be "ingested."[5] This typically happens when it is received from production (or whoever sent the video to the streaming video service). Typically the files that are received are called mezzanine files,[6] and are very large, high-quality videos that are too large to stream over the Internet. In order to prepare them for streaming, the files will go through some sort of quality check to make sure that they will look good to end users, and then will be transcoded to a format appropriate for streaming. Transcoding can change the format of the video in a number of different ways, such as changing the resolution, changing the type of compression, or changing the degree of compression. Resolution is the most common form of transcoding, which is, for example, changing a video from the ultra high-definition format mezzanine file to a high-definition or standard-definition file that can be viewed on most televisions. The file may also change compression types[7] – a very technical detail that may have significant impact on how much bandwidth is required, even for the same resolution video. Examples of different compression types include MPEG2, h.264, and ProRes 422 among others. Each different compression format has different parameters that can be modified to determine the quality and the size of the video. As a general rule, for any given compression format, the lower the quality of the video the lower the size.

Reducing the size of the video is very important for streaming over the Internet, where bandwidth can vary greatly depending on where and how a user is streaming the video. Home broadband connections can offer large amounts of bandwidth, whereas Internet connections through a mobile phone may offer very small amounts of bandwidth. To account for this, videos are offered to users in multiple sizes and qualities. Users that have lots of bandwidth available to them can watch the HD version of a video, whereas users that don't have much bandwidth can watch a very low quality version of the video. The videos are made available in a wide range of sizes and qualities to try to offer the best quality of video for however much bandwidth is available.

Part of the preparation of a video file is "chunking,"[8] which cuts the video into short pieces – such as cutting an hour long video into hundreds of two-second chunks. The purpose of chunking is to prepare the video for adaptive bitrate streaming.

Adaptive Bitrate Streaming

The amount of bandwidth available isn't always the same, even for the same person on the same device at any given point in time. For a mobile device, the viewer may be moving around and the available bandwidth may go up or down depending on the location of the viewer. For a home network, the viewer may be sharing his or her bandwidth with other people and devices in the home. If someone begins downloading a large file or streaming a video, the viewer may have much less bandwidth available than (s)he started with. The opposite may be true as well – if someone stops a download or stops streaming a video, more bandwidth may become available and the viewer might be able to watch a higher quality version of a video.

If a user's available bandwidth drops but the size and quality of the video remains the same, the video will stop playing while the video player waits for the video to download. On the other hand, if more bandwidth becomes available but the user keeps watching the same size and quality of video, (s)he might not get the best experience possible.

To account for these scenarios, one of the key technologies behind Television In The Cloud is adaptive bitrate streaming, also known as HTTP Live Streaming (HLS)[9] or Dynamic Adaptive Streaming over HTTP (MPEG-DASH).[10] This technology adapts the quality and size of the video to the amount of bandwidth available. Every video is made available in different qualities and sizes in a series of chunks. If a user starts watching the video with lots of bandwidth, (s)he will get the highest quality chunk which may be two to thirty seconds of high-quality video (the number of seconds in a chunk is something that is set at the discretion of each service). If at the end of that chunk, the video player detects that the bandwidth has dropped, it requests a smaller size and lesser quality chunk. If the video player has determined that the bandwidth has increased, it can request a larger size and higher quality chunk. This gives the video player the ability to adjust the quality to best fit the bandwidth every few seconds, ensuring the best possible user experience.

Content Distribution Networks

One of the challenges of the Internet is that users may be anywhere in the world trying to access the same file. If a user in Australia is trying to access a video in the United States, it takes a long time for the file to travel that physical distance. To help with this problem, copies of files are placed on servers that are closest to the users – in this case, the chunks that make up the video files. When a user requests a chunk of a video, the system determines which server is physically closest to him or her and makes sure that the viewer gets a copy of the chunk from that video server rather than one that is farther away. This is called a Content Distribution Network (CDN).[11] CDN services are provided by companies such as Akamai and Limelight.

Browsing and Discovery

Let's change perspectives for a moment to where things start for the end user. Initially, users will start by browsing through a catalog of content. This may be done by selecting a category such as "New Releases", a genre such as "Comedy", a favorite actor, director, price point, rating, or other information. Users may also search for content by typing in a word or phrase that they are looking for. The result of browsing and searching is typically a list of metadata that describes some shows. The metadata includes the cover art for the show, the title, synopsis, genre, cast lists, etc. It also includes some information that the user doesn't see – most importantly, the URL that tells where the video is located. The URL is exactly the same as the "http://" that a person types into a Web browser on his or her computer, but instead of bringing up a webpage, the metadata points to a manifest file. The manifest file, which is created as part of the ingest and the chunking process, describes the list of chunks that are available for that video. When a user selects a file to watch, (s)he downloads the manifest file and starts downloading the chunks to watch the movie.

Reassembly, Buffering and Rendering

At this point, the video player has the easy part of the job. It keeps measuring the bandwidth and downloading the highest quality chunk that it can. One important aspect of the video player is the video buffer. The video buffer is the video file that has been downloaded, but hasn't yet been played. If the buffer gets empty, the video player runs out of video to play and the video will either freeze or the screen will go black – not exactly what users want when they are watching the video. The biggest responsibility of the video streaming player is to make sure that the buffer is never empty. It does this by having a "low water mark"[12] – typically a size of how much video it wants to keep in the buffer. If the low water mark is two megabytes, it is the video player's responsibility to make sure that there is never less than two megabytes of video in the buffer. If the buffer drops below two megabytes, the video player will start downloading smaller and lower quality chunks of video until it can return to the low water mark.

Because the buffer is basically all the chunks put back together in the right order, at this point it is up to the rendering hardware in the video player to show the video on the screen. This is just a matter of telling the chip in the video player where the video is stored and telling it to start playing (for the most part).

Advertising Insertion

While streaming a video, it may be necessary to insert an advertisement from time to time. This is typically done in one of two ways: the server can insert the video[13] and stream it to the video player, or the server can tell the video player that it is time to insert a video and the video player will go to an advertising server to get an advertisement.[14] While inserting an advertisement on the server side is more reliable (and more trustworthy, ensuring that a viewer must watch the advertisement before (s)he receives the next chunk of the show being watched), inserting an advertisement on the video player side enables more targeting.

One of the most important aspects of advertising insertion is trying to get the most relevant advertisement in front of the right person. In the traditional world of broadcast television, everyone in the audience saw the same advertisement. With video streaming over the Internet, now every person can get the advertisement that is most influential to him or her – mothers can get advertisements for formula and fathers can get advertisements for beer (apologies for the stereotyping, but that is the purpose of targeted advertising).

Typically, the video player will know more about a user than the server. For example, a mobile phone will know where the owner is, who the contacts are in the phone, and other information that may be useful in figuring out which advertisement is most relevant to the owner. If an advertiser can figure out that the person is near a restaurant, the advertiser might send the mobile phone owner an advertisement announcing the restaurant's latest special. Likewise, if the mobile phone owner is sitting in a football stadium with 30,000 other fans, an advertiser might send him or her information about the latest sports merchandise. In order to ensure that video players are honest about who is watching the video, they are typically certified with a standard such as the Interactive Advertising Bureau's (IAB) Video Ad Serving Template (VAST).[15]

Audience Trends

The entire television industry is a delicate balancing act between content owners and audiences.[16] Content owners need audiences to monetize their content, and audiences desire premium content. The content owners are always looking for ways to make as much money as possible from their content by adding new distribution windows, increasing the number of ads, raising prices, or using any number of other tools in their toolbox. At the same time, audiences are willing to put up with quite a bit to get to the content that meets their needs and gives them the gratification they desire. However, they are always willing to pursue other forms of entertainment if the cost becomes too high.

Luckily, Television In The Cloud is beginning to offer new ways to monetize content without straining the relationship between content owners and audiences. These new monetization possibilities are important to the television industry in that it is not easy keeping audiences happy, with new screens popping up everywhere[17] and an increasing number of crowd-funded independent productions[18] available for the viewer to enjoy.

Before looking at new opportunities for monetization, it is important to look at audience expectations. As mentioned previously, the days of watching scheduled programming are slowly in decline. Although the majority of viewers still watch scheduled broadcast television shows as they air, it is expected that consumers will increasingly move to on-demand viewing over the coming years. To some extent this trend is already taking place – Netflix and their 44 million subscribers[19] is now larger than the largest cable company, which is Comcast, with their 22 million subscribers.[20]

On demand content is leading to other interesting trends. Audiences are becoming increasingly fragmented and watching more content that is more specific to their general interests, as demonstrated by the 500+ cable channels and 500 million YouTube channels. Speculation is remaining high about "cord cutting" – the act of cancelling a cable subscription in favor of watching television through other means – and binge watching is on the rise, as audiences will watch an entire season in a matter of weeks rather than over the course of several months.

However, it's not just audience behavior that is changing – the make-up of the audience itself is changing. Digital distribution makes it easier to reach global audiences, and all major broadcasters are now targeting international growth as a major part of their strategies. All this is happening across more devices than ever, as television moves from the big screen to smartphones, tablets, laptops, cars, airplanes, and beyond.

Luckily, Television In The Cloud is as much an answer to these new trends as it is a cause of many of them. Broadcasters can increasingly use their digital supply chain to reach consumers on any device in any country, enabling them to aggregate substantial audiences that have increasingly specific tastes. This is largely enabled by having web services that offer content discovery and video streaming independent of where users are or on which devices or business models they choose to watch television.

There are currently two things holding back audiences from moving to the cloud for their television viewing: (1) the availability of premium content (partially due to weak advertising monetization), and (2) a cultural trend towards continuing to use the medium they have always known. There is, however, one bright spot – mobile, which appeals greatly to the next generation and also captures new discretionary free time for watching video. Previously, users had to be in front of a television set to watch shows. Today, television audiences can watch shows while they are on a subway, out to lunch, or taking a break at work.

Monetization

Since the 1980s, broadcasters have enjoyed getting paid twice for each first airing of a television show – first by advertisers, who pay them for broadcasting commercials; and second by cable companies, that pay either licensing fees or carriage fees.[21] This economic model is partially responsible for holding back the audience demand for watching content online. If broadcasters were to show premium, first-run content online they would most likely have to forgo the licensing and carriage fees that make up half their revenue. Likewise, advertising for the same show online is typically substantially less expensive than advertising during a broadcast show, meaning that broadcasters would make less money online from advertising as well.

There is a bright spot and a promising future for being able to monetize content online. Not only does global reach mean that broadcasters can increase their overall audience size, but, for the first time, broadcasters can know something more than an educated guess about their audience. Currently, A. C. Nielsen is the source of truth for both broadcasters and advertisers for determining how many people watched a television show. Because advertisers are interested in reaching the largest number of people (and the right people) they will pay more for bigger audiences, making Nielsen ratings a critical factor in monetizing content. Those from outside the television industry are commonly shocked to learn that Nielsen doesn't actually have any way to measure exactly how many people watched a television show. In the past, Nielsen has recorded the viewing habits of around 200,000 homes[22] (around 0.0006% of the population of the United States), both through written journals of the people living in those homes that are working with Nielsen and through special set-top boxes. Those 200,000 homes aren't chosen at random, and Nielsen uses statistics to extrapolate how many people across the entire United States watched any given show – not too dissimilar to voting polls during a presidential race that are used to predict the outcome of a race.

This is largely due to the fact that broadcast is a one-way medium – the signal goes from the broadcaster to the audiences, but there is no way for the audiences to communicate back to broadcasters. Today, however, that model has changed. With cloud-based distribution, broadcasters can now understand who is watching their shows, where they live, aggregate their viewing habits, and even cross-reference their information against other purchasing and online behaviors. Advertisers are interested in reaching not the biggest audience, but the right people at the right time to influence their decision-making, and to be able to influence the way they spend their money. If someone is watching lots of television shows about babies and his or her credit card information shows that (s)he has been buying lots of diapers, it is a good chance that there is a new baby in the family, and it would be a good time to advertise formula and baby clothing to the viewer. Advertisers will pay top dollar for being able to reach the right person at the right time, meaning that as television moves to the cloud there is hope that advertising dollars alone will be able to support the monetization of content.

The ability to reach specific, targeted audiences requires content providers to invest in new cloud-based systems. These systems gather information about their consumers and compile it into a good working knowledge of who is watching (or will be watching) any given show. Having the right information aids the content providers in working with the right advertisers to buy the correct advertising time to make sure the right advertisement ends up in front of the right person. This change in advertising is no small feat and will likely take years to achieve, but it provides more value to everyone involved – higher monetization for broadcasters, the right audience for advertisers, and more relevant commercials for the audience (along with their "free" content).

Initial investments being made by broadcast networks today include substantial analytics platforms that pull in information from websites, mobile applications, Nielsen ratings, website accounts, even theme parks and credit cards. This information is used not just for advertising, but to generally understand the audiences that are watching shows. Keep in mind that broadcasters are required to pay millions of dollars up front to license content that they can broadcast, but there is no guarantee that audiences will like any given show. The result is a multi-million dollar gamble based on the gut reactions of people watching the shows and perhaps some initial research. By having deeper analytics and insights into their audiences, the content providers can understand not only what products they are interested in, but have a much deeper understanding of what the audiences are looking for in their shows. Do they like magic? Sarcastic doctors? Long twisted plots? Scandalous love triangles? Based on what broadcasters learn from their new cloud relationship with audiences, they can actually select better shows to put in front of them.

Summary

As discussed in this chapter, television in the future will be very different from how it has been over the last fifty years. Not only are business models and windowing changing, but distributors are developing an internal cloud that will be used for a digital supply chain and to implement automation through a file-based workflow. Some of the changes are already apparent, with streaming video technologies already becoming available and widely implemented. How the future will develop will depend on the audience preferences that are beginning to be observable in today's trends, and through the different mechanisms for monetization that content owners have now and in the future.

CHAPTER 7
THE CLOUD AND TELEVISION

While the opening section of the book focused on the various aspects of cloud computing in general – the types of clouds, the different cloud service models and the advantages of using the various types of cloud computing, the rest of this book focuses on how Television In The Cloud works and the importance of the cloud for television.

As has been stated before – and will be stated again – television is made for the cloud and the cloud is made for television. Television In The Cloud will be the cornerstone of what will become the next revolution in television – a revolution so transformative that the entire industry will be rebuilt. This transformation will usher in a new era in the television experience for both the industries and the television viewers. Television is made for the cloud, the cloud is made for television, and television will never be the same.

Welcome to 21st Century Television

Of all the different technologies – both legacy and new – Television In The Cloud technology may be the most important technology to underlie, and continue to underlie for the foreseeable future, television in the 21st Century. Cloud computing technology makes it possible for content providers (the networks, the cable channels, the myriad online channels, Netflix, Hulu+, YouTube, etc.) both to store their inventories of programming and to make it possible for their viewers to access that inventory anywhere, anytime, and on any device. Television In The Cloud technology makes it possible for the content providers to have a way to bring their program inventories together in one location so that viewers can easily retrieve and enjoy those inventories. Television In The Cloud provides viewers the opportunity to enjoy the content providers' inventories in a variety of ways using the providers' web sites or television apps and widgets. Whether they are watching the programming in their homes on their connected television sets, on the go using their smartphones or tablets, or multitasking using a combination of technologies, cloud technology is what makes Television In The Cloud – 21st Century Television – possible.

In his book, *21st Century Television: The Players, The Viewers, The Money*, the author, Dr. Frank Aycock, described Television In The Cloud this way:

> Visualize for a moment 21st Century Television as cloud television. Under this scenario, IPTV becomes the backbone connecting the viewer with a myriad of television choices in the cloud, or more properly, in a host of clouds, whether by wired connection, Wi-Fi, cellular, or some other types of connections yet to be developed. Through the use of IPTV and cloud television, content providers ranging from the traditional broadcast and cable/satellite providers to today's alternative providers delivered through set-top boxes and over connected television sets using software from Google or others, to the separate streamers such as Netflix, Vudu, and Hulu, among others, to the user generated content sites such as YouTube and SocialCam, reach their audiences with any program or content they desire, anywhere, anytime, and on any platform.[1]

Television In The Cloud technology is what makes the future of television in the 21st Century and beyond possible.

Because the content providers will have their entire inventories of programs and content in the cloud, they will be able to make any episode of any program in that inventory available to a viewer in a perpetual video-on-demand format. Gone will be the need for determining which few programs will stay on the air and available for the viewers to watch. Television In The Cloud provides a safe, secure, cost-effective, environmentally-friendly way of providing programming to viewers – and providing the viewers with the programs they want to see, when they want to see them, instead of forcing them to watch (or not watch) programs they may not care about, find ridiculous or offensive, or are not in their frames of interest.

For the content provider, Television In The Cloud makes connecting with the viewer quick and easy. When all television is in the cloud, programs can be accessed at anytime, anywhere, and on any platform. Whether viewers choose to watch their favorite programs on their big-screen ultra-high-definition, 3-D, holographic television sets, or on their computers, their tablets, or their smartphones, Television In The Cloud makes it easy and simple to watch. Further, the content providers can make it possible for their viewers to be able to choose to watch their favorite programs anywhere in the world that there is an Internet signal - not just in the country of the program origin, or a country where the program is telecast in a traditional manner.

For viewers, Television In The Cloud is the basis for true TV Everywhere. As mentioned in the previous paragraphs, Television In The Cloud makes ubiquitous viewing available to anyone as long as there is Internet capability available, whether it is through a wire, through Wi-Fi, through a cellular connection, or through some other form of delivery. Further, combining refined search and promotion[2] with IPTV delivery of programming through apps, widgets, or other forms of direct connections makes Television In The Cloud the linchpin of 21st Century Television.

Television In The Cloud's Impact on The Television Industry

Television In The Cloud will have a major impact on all the television industries from the legacy industries to the new media industries. While Television In The Cloud's impact will be felt in positive ways for many of the different parts of the television industries, for others the impact will likely be negative. The first book in this series, *21st Century Television: The Players, The Viewers, The Money*, describes the future of the various media industries, both legacy and new. For those who have not read the first book, this chapter will weave into the discussion an understanding of the future of the various television industries without trying to repeat what has already been discussed.[3]

The Legacy Industries

For the legacy television industries, the impact of Television In The Cloud will be mixed. Many of the positive impacts are discussed in chapters later in this book. Overall, the broadcast networks are facing major changes and Television In The Cloud will be at the heart of those changes. In the future, the broadcast networks will no longer deliver their programming through their current local affiliate stations. Rather, they will drop their affiliates for the possibility of delivering their programming directly to the viewing public. Television In The Cloud makes that move possible, because – as mentioned earlier – the technology allows the networks to move all their programming to the cloud. Moving their entire inventories to the cloud lets the networks make their programming available 24/7/365 to their audiences. Not only that, but it also makes the programming available to the viewers in an "anywhere, anytime, on any platform" position.

Additionally, moving their programming to the cloud to allow their audiences to watch their programming makes it possible for the networks to bypass organizations such as Aereo and other over-the-top delivery technology. Further, Television In The Cloud makes it possible for the networks to develop new programming contracts with the pay-TV companies for the right to carry their cloud-based programming over the pay-TV companies' broadband connections. Additionally, the networks – because their programming would be in the cloud – could also require subscriptions from viewers for the right to view their programs along with added bonuses of making the subscriptions more price-worthy. The combination of pay-TV contracts and viewer subscriptions easily should more than offset the current retransmission contracts the networks have to negotiate today.

Further, Television In The Cloud makes it possible for the networks to advertise directly to the viewer using addressable, customizable advertisements, delivering products and services to viewers with the messages and creatives that will stimulate the viewers to respond positively. While true, individual microadvertisements are neither cost-effective nor feasible, strategies such as Aggregated Targeted Microadvertising, or ATMA, make addressable advertising feasible, cost-effective, and, generally, highly successful. Strategies such as ATMA also allow the network to sell an individual availability several times, thus more than compensating for the additional cost of production and smaller audience for each of the several advertisements aired to reach specified aggregated viewers. Further, because of the highly targeted nature of each advertisement during the availability slot, the likelihood of a successful end result is very high as well, making the availability timeslot even more valuable and, therefore, more expensive. In total, the networks stand to make, at worst, the same amount as they currently do and, at best, much more revenue than today. Television In The Cloud also opens the door to advertising to binge watchers, those viewers who prefer to watch a season or even a series in one or a few sittings.[4]

Finally, using Television In The Cloud, the networks could provide their programming on a global basis, attracting audiences and advertisers from around the world to their programming. The global nature of Television In The Cloud makes it possible to reach huge untapped audiences in countries across the world with programming that they would want to see (U.S. television programming in syndication is already the most popular worldwide), in a way that allows those viewers to watch at the location and in the manner of their choosing. Additionally, Television In The Cloud makes it possible for the networks to reach their U.S. audiences when they travel abroad as well with the programming they want to watch and advertisements for products they would like or that they might find of value that are local to the countries the viewers are in.[5]

Turning to cable, for the cable channels such as ESPN, HGTV, USA, etc., Television In The Cloud will offer the same possibilities that it does for the broadcast networks in reaching its current viewers, developing more successful highly targeted advertising, and attracting new audiences from around the world. Additionally, because all television will be delivered from the cloud to the viewer through IPTV,[6] the cable channels will be on a more equal footing with the broadcast networks in terms of attracting advertisers and audiences. Television In The Cloud is a truly positive opportunity for the cable channels in the future.

For the companies owning the cable systems around the U.S., Television In The Cloud will cause what could be a seismic shift in the way the cable companies deliver programming to their subscribers – or maybe not. All the major cable companies today offer broadband subscriptions to their subscribers, often at speeds greater than subscribers can get elsewhere. A move by the broadcast networks and the cable channels to Television In The Cloud would allow the cable companies to eliminate their traditional cable hardware and transition to high-speed Internet connections to all their subscribers' homes. In addition to shifting the cable companies' focus to high-speed broadband delivery solely, Television In The Cloud would allow the cable companies to renegotiate their carriage fees contracts and to re-assert themselves in the delivery of television to the homes in the U.S.

Additionally, Television In The Cloud will provide the impetus and the means for the cable companies to make "TV Everywhere" a reality. Further, as the cable companies continue to consolidate, the likelihood of the cable companies to expand their broadband abilities nationwide will become a reality. Instead of local monopolies, the combined cable companies will compete head-to-head for audiences not only throughout the U.S., but also around the world as cable companies around the world consolidate to reach worldwide audiences as well. Television In The Cloud, then, becomes the major driving force for the cable companies to grow exponentially into worldwide television powerhouses.

One quick mention in this section on cable is to quickly comment on the two major telecom competitors to the cable operators – AT&T's U-verse and Verizon's FiOS. Television In The Cloud offers both companies the same opportunities and advantages that it does to the cable companies for growth and expansion. As to how and if these two competitors will take advantage of the possibilities offered by Television In The Cloud remains to be seen. It is possible that both could be swallowed up by the behemoth global cable companies of the future, turning them into the telecom arms of their respective corporations.

The New Media Industries

The new media industries likewise will be impacted by Television In The Cloud. Many of the new media technologies have already chosen cloud-based technologies to store and deliver their programming to viewers. However, as Television In The Cloud becomes ubiquitous, a variety of those new media industries could experience a negative impact, as well as the positive impact that Television In The Cloud has had in their development.

For Internet Protocol Television, or IPTV, Television In The Cloud will be the overlaid technology that IPTV will deliver. As 21st Century Television comes to its fullest fruition, IPTV will be the delivery mechanism for all of television. Already AT&T and Verizon are delivering their pay television services using IPTV. Cable companies are moving to strengthen their IPTV services to be prepared for the day when all of television is delivered through IPTV. As far back as 2010 (practically a lifetime for technology these days), cable executives were saying that the future of cable is not cable – it's broadband[7] with enough downstream speed to make it possible to deliver IPTV through a variety of technologies, both wired and wireless. Internet Protocol Television will make it possible for all IPTV deliverers – cable or otherwise – to bring Television In The Cloud programming to viewers no matter where they are. The possibility of TV Everywhere lies in the combination of Television In The Cloud being delivered by IPTV.

Television In The Cloud also makes possible complete video-on-demand, although perhaps the more correct term should now be television-on-demand.[8] The days of a network or cable channel programmer making the decisions for the viewers as to what they will watch and when they will watch it are numbered. Appointment viewing is dying – television-on-demand is the future of television. When networks and cable channels place their entire inventories onto the cloud, viewers can have complete control over what, where, when, and how they watch their favorite programs. No more will the viewer have to wait until a certain day and time to watch his/her favorite program – Television In The Cloud makes watching a favorite program completely at the discretion of the viewer. It is the viewer that will make the choice of the program to watch, when to watch it, where to watch it, and how (s)he wishes to watch it. Television In The Cloud also makes it possible – through television-on-demand – for a viewer to watch a favorite episode multiple times back-to-back, and makes it possible for the viewer to watch an entire season or series of a program at a single sitting, or over the span of a few sittings. Netflix and Hulu/Hulu+ are but two examples of early television-on-demand driven by Television In The Cloud.

Additionally, television-on-demand driven by Television In The Cloud makes it possible for the networks and cable channels to promote their entire inventory of programs on a constant basis, because there is no need for a program ever to disappear from the content provider's inventory. Programs that are no longer "current" in today's way of thinking will be just as viable and just as deliverable to the viewers as they were when they were on the current lineup of programs. Additionally, there is no longer a need for schedules other than to announce when a live event will take place – the date and beginning time of the event being all that will be needed – or the initial running date and time of a new program just being introduced by the content provider. Further, the only reason for the announcements would be for those viewers who want to watch the live event in real time or who always want to watch a new program exactly when it first airs. Over time, however, given the changing viewing habits of the younger generations, the likelihood of sizable audiences wanting to watch a new program exactly when the first episode initially airs will grow smaller and smaller. Television In The Cloud delivering television-on-demand to viewers demanding the right to control their own viewing will be painful to the legacy media content providers, but it is inevitable that the day will come.

For the traditional set top boxes, or STBs, Television In The Cloud holds great promise in the short and probably medium run, but ultimately holds the death of the STB in the long run. STBs have had a long and important history in the development of television.[9] Even today, set top boxes from Roku, Apple TV, and other stand-alone boxes; to the videogame console; to the Blu-ray digital video player and digital video recorder; to the TiVo and other boxes provided by the cable and DTH satellite companies are important – often critical in the minds of the owners – additions to their television hardware choices. Many of the companies that provide the set-top boxes use the cloud for their program delivery to great effect. Of those program delivery services, Netflix and Hulu/Hulu+ are the best known. In the short run, television viewers will continue to enjoy the different STBs that are available to make viewing of programming from both the legacy media and new media choices possible.

However, in the long run, Television In The Cloud will not be limited to the set top box. Even today, the first uses of connected television sets, tablets, and smartphones to view television are being enjoyed by viewers throughout the country and around the world. Those set top boxes that do survive past the short and medium-runs will be the boxes that will serve more as "gateways" to the cloud. These gateways will deliver programming to the television set while at the same time being the deliverer of wireless services to the home computers, tablets, smartphones, and other services around the home such as smart refrigerators, smart ovens, smart heating and air conditioning, and other such services. Without the need for the ancillary services, the connected television set with the ability to deliver programming to viewers will make the set top box obsolete and it will have the ability also to deliver the wireless Internet connection that the home computers, tablets, and smartphones will need. In the long run, the venerable set-top box will be a transitional device, one that will be destined for the museums of television around the world.

While Mobile DTV is, or will be, dead in the next year or so, a successor to Mobile DTV could be an answer to video viewing while on the go.[10] However, it will not be the broadcast of local signals to television sets and smartphones in automobiles when the passengers are within the local market. Rather, the successor to Mobile DTV will be a form of Television In The Cloud delivered as Internet TV. Already today, it is possible to receive direct-to-automobile and direct-to-home unit satellite radio from Sirius/XM, and Internet radio stations from Pandora and others in the automobile while on the go. Moving from the delivery of Internet radio to Internet television – delivered, of course, by Television In The Cloud – is the obvious and most cost-effective move for Mobile DTV (or whatever name the industry wishes to use). Television In The Cloud makes television viewing possible anywhere there is an Internet connection, so watching television on the go is a given accomplishment in the near future.

Television In The Cloud is the driving force that makes the future bright for the connected television set. The connected television set, by its name, is a television set that is "connected" to the Internet through a wired connection, a wireless connection – either Wi-Fi or, in the future, cellular – or some other form of connection that has yet to be developed and implemented. The connected television set is still in the early stages of consumer adoption. However, as the legacy media find it necessary to move to the Internet to reach their audiences, the connected television set will become more and more integral to the home television viewer because of its ability to deliver all aspects of the Internet, not just television. For home television set viewing (as opposed to tablet, computer, or smartphone viewing at home, or on-the-go viewing using those same items along with the successor to Mobile DTV), the connected television set will be the television of choice for viewing audiences. Television In The Cloud will make it possible for the television viewer to be able to watch his/her favorite television program, surf the World Wide Web, continually update her/his social media sites, and get real-time updates on other programs of interest, promotional announcements, and a host of other information of interest.[11]

Summary

Television is made for the cloud and the cloud is made for television. Regardless of how the viewer will watch television in the future, it will be driven by Television In The Cloud. If the viewer is at home, watching television on a huge connected television set – Television In The Cloud over IPTV and by video-on-demand will be the way the programming will be delivered to him/her. If the viewer is traveling and watching television on a tablet, a laptop/netbook/hybrid-tablet-and-computer, or a smartphone – Television In The Cloud will be there, delivering the programming the viewer wants to enjoy at the time and location (s)he wants to enjoy the programming. No matter where in the world, no matter what time of day or night, no matter the platform chosen – Television In the Cloud will be there for the viewer, making sure (s)he has the opportunity to receive the programs anywhere, anytime, and on any platform.

CHAPTER 8
MAKING USE OF TELEVISION IN THE CLOUD

The cloud is made for television and television is made for the cloud. That has already been said numerous times throughout this book. Television In The Cloud has the ability to affect every aspect of television, from local stations to the networks, to cable channels in a positive manner. This chapter will take a look at each of the different areas (departments, divisions – take your pick of terms) of television and discuss what Television In The Cloud has to offer and how it will impact each of those areas.

Programming

Television In The Cloud makes it possible for content providers – the networks, the cable channels, the local stations – to have a way of storing their inventory of programming that allows the viewer to be able to access the programming anywhere, anytime, and on any device. Additionally, Television In The Cloud technology provides content providers an excellent way to access their inventories of shows and bring them together in one location so that viewers can easily retrieve and enjoy the content providers' programs. Finally, Television In The Cloud technology gives viewers a way to enjoy the content providers' inventories using the providers' websites or television apps on their connected television sets, tablets, computers, and smartphones. Let's look at each of these points individually:

1. Storage – The cloud is an excellent storage device. Chapter 2 and Chapter 4 discuss the various types of clouds and the benefits that are derived from cloud computing. For television content providers, depending of their needs and how they wish their viewers to access the programming, any of the three types of clouds – public, hybrid, and private – can be used to store programming. Further, because storage in the cloud is (generally) easily scalable, content providers can choose the amount of storage they will originally need and scale up or down as they add new programs or remove old programs deemed no longer to be cost-effective to maintain. Additionally, because scalability is often on a pay-as-you-go system for most third-party cloud providers, the costs of storage can be effectively controlled, again making the cloud cost-effective for the content provider. The cloud also makes it easy to upload and store a program, whether the program is a scripted one that's stored for later release, or whether the program is a live event with simultaneous storage. The flexibility of the cloud makes storing television programming easier and more cost-effective than other forms of storage. Finally, storage in the cloud is safer than storing on different media such as reels of tape or on an internal server. The cloud protects the programming from decay, damage, or destruction. Overall, the cloud provides the safest and most cost-effective way to store television programming, regardless of the content provider, but especially for those providers delivering programming to a nationwide or larger audience.

2. Inventory Access – Today's content providers have a problem when programming is stored in a variety of locations and is being delivered from a number of locations as well. Some programs may be in Los Angeles, for instance, while other programs may be in New York City. Coordination between the locations is required to have a smooth-running, seamless schedule of programs, advertisements, and promotions each day. Television executives obviously are accustomed to the demands of current television coordination. However, having the ability to access all programming from the cloud makes the problem of coordination irrelevant. As television moves from appointment viewing to on-demand viewing, access moves from the television executive in charge of programming to the viewer who chooses what (s)he will watch at a given time. Access becomes whatever the viewer wishes to watch and the need of the programmer is no longer scheduling, but providing a system of search and discovery designed to acquaint and interest the viewer in the content provider's inventory of programs. The cloud makes access easy, cost-effective, and viewer-friendly.

3. Ease of Viewing – Today's viewer would like to have his or her television anytime, anywhere, any way (s)he chooses or best fits the situation. Tomorrow's viewer will *demand* no barriers to his/her viewing. The content providers of today can still draw audiences using schedules for programs and expect viewers to appointment-view those programs. The content providers of tomorrow will not be as lucky. They will be expected to cater to the viewer's wants and needs, not the other way around. Those that do will be tremendously successful; those that don't will not survive – it's that simple. Television In The Cloud makes ease of viewing possible for the viewer no matter where (s)he is in the world. Television In The Cloud makes ease of viewing possible for the viewer no matter what time (s)he wants to watch the program. Television In The Cloud makes ease of viewing possible for the viewer no matter how (s)he wants to watch – one episode at a time, an entire season in one sitting, or an entire series in one sitting or over a set period of time. Whether it be on connected television sets, tablets, smartphones, Google glasses or other wearables, bathroom mirrors that double as screens, or whatever else will be dreamed about and then developed for the viewer, Television In The Cloud will make ease of viewing a reality. With Television In The Cloud, viewers can watch their favorite programs at home, on the road, in hotel rooms, guest houses, apartments, condos, or wherever they are just by selecting a content provider's website or choosing the content provider's app. Viewer control of television is what Television In The Cloud provides, and in the future, the viewer will accept no less than complete control over his or her viewing.

Work on cloud technology to deliver programming to viewers is already occurring. Netflix, Amazon Instant Video, YouTube, Yahoo, Hulu and Hulu+ are all examples of cloud-delivered television. The CW network is streaming its programming concurrently with its over-the-air broadcasts. ABC's "WatchABC" app allows cable and satellite subscribers to watch real-time programming through their connected TV sets, tablets, and smartphones. CBS is also working to stream its content in real time using Syncbak software. Aereo and FilmOn X, both currently in litigation with the networks over copyright and retransmission consent, are forms of cloud services that provide viewers within local network affiliate stations' markets to watch over-the-air broadcasts of those stations' programming on their tablets and smartphones. The problem for the networks at this time, of course, is how to keep their affiliates happy. This will change in the coming years as the whole relationship between the networks and their local affiliates changes and evolves. Television In The Cloud will also be a catalyst for that change as well.

Production

Television production will benefit greatly from Television In The Cloud. Television In The Cloud makes it possible for local television stations, broadcast networks, cable channels, or production houses, to produce finished programs as they are shot on location instead of having to bring the recorded material to the home location for editing. Traditionally, the production of television programs, and especially, the video editing of those programs has been tied to the edit bay.[1] Television In The Cloud makes it possible for each camera used in the shooting of the program to send its footage directly to the cloud, making it instantly available to an editor either on location or back at the home location. Television In The Cloud also makes it possible to upload other raw or finished footage to the cloud, along with graphics, stills (if needed), audio, etc., from wherever the production location might be. With the combined footage available in the cloud, editors onsite can download and edit the production while the crew is still on location. Future options will include editing directly in the cloud itself.

Simultaneously, using Television In The Cloud, the editors at the home location have the ability to watch the editing as it progresses and check the incoming footage for quality. They can order changes to the on-location editing as additional footage becomes available or request more footage or different shots if needed or desired. In other words, Television In The Cloud makes it possible to control the production from start to finish in real time or, at worst, in near real time. In short, Television In The Cloud technology will reduce the time it takes to deliver finished productions – from local station or network advertisements, all the way to full length, on-location television programs.

The same will be true for in-studio programs as well. Shots from each of the cameras used for the program can be sent to the cloud where editors can select and edit the program in near real time. The ability to shoot and edit virtually simultaneously makes producing studio programs easier and quicker to prepare and more cost-effective. Each episode will be in finished form almost immediately after the production is given the "wrap," making it possible for programs to be prepared and available for the viewing audience significantly quicker than is currently possible.

Finally, Television In The Cloud makes it possible for live programs to be recorded, edited for future playback time constraints (especially when airing sporting events), and made available for audiences worldwide. While, generally, live events, due to their fact that they are "live," will likely make less use of the cloud in the early stages, recording to the cloud will make instant replays and isolation shot recalls as easy and quick as current technology, but without the concern for damage to the footage or equipment.

In summary, Television In The Cloud technology will make television production much more cost-effective and will reduce the time-to-air significantly. As Michael Worringer, the director for video products at the online video platform Broadcast Interactive Media, says, "Postproduction in the cloud is working and will proliferate. We still run into broadcasters tied into legacy workflows and hardware where they have two or three people managing servers on site costing thousands of dollars a year. Sometimes we scratch our heads. Cloud production will accelerate as the need to produce more and more content, faster and with fewer resources gets ever stronger."[2] The major companies in the television editing space, such as Avid and Adobe, as well as smaller companies such as Make.tv are developing, or have developed, cloud technology specifically designed for cloud-based production.[3]

News

Television In The Cloud technology will enhance the production of news in many of the same ways that it enhances other forms of production. Much as with the discussion of production of television programs in the section above, news stories can be shot on location and all the pieces of the story – the stand-up, interviews or soundbites, and cover shots – can be loaded onto the cloud. From there, editors at the station or network, or even on location, can assemble the parts of the story into the full package, ready for air. Further, with today's on-site editing capabilities, often the story can be edited right on the camera or smartphone or tablet computer. The completed package can then be sent to the station or network for airing. Because of the immediacy of news stories, the ease and time reduction of using Television In The Cloud to deliver finished news stories from anywhere in the world to air is extremely cost-effective. All the news crew needs is Internet access – which they can have through a mobile Wi-Fi hotspot or even cellular connectivity to deliver either the parts of the package to be assembled at the station or network, or the final edited story back to the home location.

Additionally, there is a cost saving from no longer needing dedicated transponder time on a satellite to send the story to its ground location which is substantial. As of September 2013, all of CBS' Newspath affiliates will be downloading the stories they receive from the cloud using an Internet browser instead of a satellite feed. ABC NewsOne will launch a new, totally redesigned cloud system sometime the early part of 2014, and CNN's Newsource has moved nearly all of its subscribers to a cloud-based system.[4] Unlike satellite services, Television In The Cloud does not require stations to have costly upload/download satellite dishes or internal servers to store the material being sent. Dave Cunningham, president and founder of Generation Technologies, a cloud service especially designed for news organizations says,

> It takes about 11 minutes for one station to upload a three-minute clip to the cloud, the system to transcode it into multiple formats – SD, HD, MPEG-2, and h.264 – and for another station to download it. When everyone was using satellites, especially leading into nightly newscasts, the system would back up. That same three-minute clip could take an hour to upload and download. There was a reason why, in the satellite world, we strongly encouraged that contributions be under three minutes. It just takes too much time on a satellite to transport that much content. And if it's really busy, a clip could have uplink fade, and then the broadcaster has to transmit it again.[5]

As to costs, Cunningham says, "Suddenly, you have no server, no dish, no I-band infrastructure – all of that is really expensive."[6]

ABC NewsOne's cloud solution, called VideoFusion, is expected to be up and running early in 2014. Al Prieto, Vice President of NewsOne says, "It's a complete refresh. It'll look different, work better and have more bells and whistles that makes it easier to access content in all file formats. It'll be a big improvement."[7] CNN was one of the pioneers of cloud-based solutions for its news division, beginning in 2010 with the Bitcentral Oasis system. By 2013, nearly all CNN Newsource subscribers were using the cloud system for downloading the cable network's content.[8] Fred Fourcher, founder and CEO of Bitcentral says the reason that so many of the Newsource subscribers use the cloud-based systems is that "…you can get more with the Internet download than you can with the satellite download."[9] Television In The Cloud is a growing part of the news production of television stations, broadcast networks, and cable news networks, and will continue to grow in importance in the years to come.

Sales/Advertising

It is in the area of sales and advertising that Television In The Cloud technology will likely have its most important impact. Because the cloud allows access from anywhere there is an Internet connection, on any device, at any time, cloud technology makes it possible for content providers to produce revenue through advertising and product placement in new and exciting ways. In terms of advertising, cloud technology makes it possible for content providers to deliver to the viewer addressable advertising – that is, individualized, personalized advertising for that specific viewer. Today, networks and local stations – and, to a lesser extent, cable/satellite channels – use a shotgun approach, trying to reach the largest audience and hoping that "something sticks" with viewers. Using Television In The Cloud technology, the viewer can receive an advertisement designed for him or her for a product (s)he is interested in and has a desire for already. While today's viewers typically hate watching advertising – getting up to go to the restroom or the kitchen while the advertisements are running or using the DVR to "skip" the commercials altogether – even millennial viewers will watch advertisements they want to watch for products they want to have. Making ten million different ads for ten million different viewers is not cost-effective, nor really even economically feasible. However, other alternatives such as Aggregated Targeted Microadvertising (ATMA) make it possible to group viewers through personalized information and purchasing habits and offer different ads to the different groups.[10]

Combining alternatives such as Aggregated Targeted Microadvertising with Television In The Cloud makes it possible for content providers to sell each commercial spot or network availability numerous times to a variety of different product clients depending on the different group preferences. Viewers may have different likes and dislikes in terms of products and services, but a number of different viewer groups will intersect at the individual program. Using ATMA through Television In The Cloud, different ads can be delivered to each of the different viewer groups simultaneously, satisfying each group's needs, wants, and desires for their preferred products and services. While each advertisement may not be able to be sold for as much money as before because it would be reaching a smaller audience group, there could easily be no drop-off in price given the almost certain success rate of the ad. Further, because the ad time would be sold several times, the total revenue generated for each ad could – and would – likely be significantly higher than today. Companies such as BlackArrow and YuMe are just a couple of the numerous companies already providing the first generation of software to make addressable advertising possible.[11]

Television In The Cloud also provides a second avenue for significant revenue generation through ubiquitous product placement. With Television In The Cloud technology it is possible to make every element in a program a clickable link. A viewer can see an automobile in the background that (s)he likes. By clicking on the image, the viewer can find out what make and model it is, and where (s)he can get more information, purchase the car locally, etc. Other such examples could be finding out more about the Apple computer the star is using just by clicking on the product; or noticing an interesting watch, piece of jewelry, or article of clothing an actor is wearing and clicking on the item to find out who makes the product, information on the quality of the product, where to buy the product either locally, or, if online, to buy the product immediately. Television In The Cloud takes product placement to a whole new level of viewer involvement, and one that most viewers and especially younger, more tech-savvy viewers will appreciate and involve themselves with regularly.[12]

Promotion

As Television In The Cloud impacts programming in the future, promotion becomes critical to the success of the content providers. However, promotion also must encompass an optimized search capability that will lead (or drive) viewers to the content provider's inventory of programs. While the programming departments of the content providers may find it necessary to shrink in size, the increase in the promotion department's staff will more than offset the loss of positions in the programming department.

While many consider promotion and social media to be equivalent, in reality, social media is only one part of the entire promotional effort of a content provider. The revolution that will be the 21st Century Television universe will include the following promotional efforts:[13]

1. Search – Because Television In The Cloud makes it possible – and even most probably required – for a content provider to have its entire inventory of programs available for viewing at any time 24/7/365, search becomes paramount in finding and delivering the programs to the viewers that they want to watch. For Television In The Cloud, search becomes a structure for finding the right program for the right viewer. Search will include not only traditional search engines, but also new, specialized search engines designed and operated specifically for the television industries. Viewers looking for programs will go to these specialized search engines to find programs based on factors including (but not limited to) name of program, genre, content provider, the name of the star or stars, or the name of any other supporting cast member. These specialized search engines will allow the viewer to quickly and easily filter to the program (s)he wishes to watch.

2. Website Promotion – With Television In The Cloud, all content providers will be forced to maintain their websites and any additional specialized websites they may use specifically for various television content (most do today anyway). In the future, the website(s) will most often be the "front door" through which viewers find information on the content provider's inventory of programs, dates and times of live events, and new programs being introduced. Maintaining the website and making it as easy as possible for the viewer to find and watch the programming (s)he desires will also be the domain of the promotion department.

3. Social Media – Of the different areas, more has been written and studied about social media and the various ways it impacts television today. Television In The Cloud makes social media engagement an intensive and continual exercise in connecting with and persuading viewers to choose to watch the content provider's programming. Employees of the promotions department of a content provider will have to be adept in making full use of the major social media sites, including Facebook, Twitter, Pinterest, etc. Most likely every promotions department will have a section of employees devoted to nothing but keeping the chat flowing on all the social media sites. Others will be responsible for making sure the cast and especially the stars of the popular and the new programs just released or close to release are filling their social media sites with news about themselves and the programs they are or will be a part of, so that the viewers can connect with the stars and the programs.

4. Other Online Sites – Promotion in the age of Television In The Cloud will require making promotional video clips available to other online sites on a continual basis. Promotional clips of current, classic, and newly released programs; of upcoming live events; and of scripted programs to be released in the future will be one of the major promotional tools available to the content provider. In addition, interviews with cast members and especially the stars of the programs, along with promotional spots of the stars and cast members extolling the positive aspects and enjoyment of their programs will air continuously on such sites as YouTube and others.

While promotion has always been an integral part of television content providers, the requirements of Television In The Cloud make the promotion of a content provider's entire inventory of programs crucial to the short and long-term viability and profitability of the provider.

Global Delivery

While not an area of consideration in today's television industry, Television In The Cloud makes it possible for content providers to deliver programming to audiences throughout the world. The capability to deliver television programming globally is and will be the most complex and controversial of the various uses for

Television In The Cloud technology. While some would suggest that the engineering aspects of Television In The Cloud will be the most challenging, in reality, those holding the view are wrong. The technology to deliver Television In The Cloud – from an engineering standpoint – is developing quickly and will be available to handle virtually any type of television program available within the next few years. Rather, the major problems that are needed to be overcome for Television In The Cloud to become viable are carriage agreements and copyright requirements, both of which are less about technical aspects and much more about negotiation and cooperation. However, IPTV is the fastest growing sector of television delivery today in both the U.S. and especially around the world. Already, companies such as Netflix are finding ways to extend their reach into other countries, and certainly YouTube delivers a huge assortment of user-generated and other, more professional-quality videos worldwide from producers who are worldwide. While Netflix and YouTube are highly specialized companies doing highly specialized operations, they demonstrate the possibilities that are available to the content provider and to the viewer.

What is ultimately required, then, is the willingness of whichever parties are involved to cooperate in the negotiations over copyrights and carriage requirements. Content producers and those holding the copyrights to programming must be willing to negotiate with the content providers to deliver their programming that include the rights to worldwide distribution of the programming. Both the content providers and the content producers must be willing to negotiate in good faith the worldwide rights to the programs.

Second, the various countries around the world must be willing to accept and to embrace the notion of Television In The Cloud as the global delivery mechanism it is. No longer will governments be able to place restrictions or quotas on television when it is delivered using the cloud without blocking those portions of the Internet where the content providers are located. To do so would, in all likelihood, fail in most countries due to the global community that is the Internet. The positive side, though, is that – as television becomes more global - countries will be able to receive tax revenues from those global content providers. The opportunities for tax revenues include everything from income taxes paid by employees of the global content providers to corporate taxes for offices in the specific country to the traditional taxes required of television content providers and television advertisers in countries today. In reality, the success of Television In The Cloud as a global system boils down to negotiations. With the promise of global audiences and worldwide advertising revenues, it is likely that the success of Television In The Cloud development and implementation in the reasonably near future is virtually certain.[14]

Summary

Television In The Cloud technology is in a state of rapid development and improvement. In virtually every area of the television business, Television In The Cloud is having a transformative impact. But Television In The Cloud is a technology for both the present and the future. Even more than the overall cloud computing, it is in the relatively early stages of development. To be sure, Television In The Cloud technology is being deployed today in an increasing number of ways by the legacy content providers, and is in use by the new media content providers.

However, much more can and will be done with Television In The Cloud in the future. As the rampant changes in television technology continue to develop and grow, Television In The Cloud will play an ever-increasing role in the continuing development and implementation of 21st Century Television. The cloud is made for television and television is made for the cloud. There is no better proof of that than Television In The Cloud.

CHAPTER 9
THE FREEDOMS OF TELEVISION IN THE CLOUD

(AUTHOR'S NOTE: *This chapter is an extended adaption of a TED talk delivered by Dr. Frank Aycock on September 7, 2013, at the TEDx Conference in Nagoya, Japan, and a keynote presentation at the Cloud Computing Conference – West in Las Vegas, NV, on October 28, 2013.*)

Television and the cloud are made for one another. Let me say that again – television and the cloud are made for one another. As has been shown in the previous chapters of this book, the question can no longer be "if" television will move to the cloud, but "when" Television In The Cloud will be fully implemented throughout the United States. If you have read this far in the book, and you are an engineer, computer science person, IT person, or some other type of professional that expected the authors to reveal the magical new formula that will make it easy, practical, and cost-effective to make Television In The Cloud work – well, that is obviously not going to happen. It will be up to you or others like yourself to develop the methods and technologies necessary to bring Television In The Cloud to complete fruition. Even this late chapter will not tell you "how" to make Television In The Cloud happen; rather, its purpose is to tell you, the reader, "why" the move of television to the cloud must and will be made – and sooner rather than later. Let me say it again - television and the cloud are made for each other.

Twenty-first Century Television – the future of television, if you prefer – is, at its core, fundamentally about freedom – freedom for both the television audiences and the television industries. That view has been well-stated in the book, *21ˢᵗ Century Television: The Players, The Viewers, The Money*. These freedoms will, in large part, be driven by Television In The Cloud. If you have a problem with the word freedom, think of those freedoms as major benefits.

Today, when someone begins talking about Television In The Cloud, (s)he must realize that cloud-based television is a controversial subject. There are numerous problems and concerns to consider when looking at Television In The Cloud, including engineering problems, political problems, social problems, legal problems – yet, none of these problems or concerns are insurmountable – they can, and will be, overcome. Further, it is important to remember that the development of Television In The Cloud is not something that one can expect to happen within a week, the next month, or even the next year from when the reader finishes the book. Rather, it is extremely important for the reader to consider the possibility of Television In The Cloud reaching completion somewhere around the year 2025, eleven years from the time of publication of this book. It perhaps might even be preferable to consider 2020, six years from now, as the completion date – primarily because the speed of technological innovations today, especially in technology needed for Television In The Cloud, can easily make one an optimist, and lead one to believe 2020 is likely more realistic.

In today's technology universe as well as today's television universe, 11 years, or even six years, is a lifetime, or maybe even several lifetimes. Think back to 11 years ago – 2003 – then think of what has transpired in both technology and television since that time. The developments have been astonishing. It is extremely doubtful that most people – engineers or not – would have considered it possible that back in 2003 they would be talking about Television In The Cloud just eleven years later in 2014. Quite likely, if someone had mentioned cloud back in 2003, most people would have looked up into the sky and asked which type – cirrus, stratus, cumulus, or nimbus? After all, the changeover from analog to digital television – in the U.S. – was still six long years away. There was no iPhone, no iPad, no app stores, no consumer-available HDTVs – at least in the U.S. – no 3D TV sets, no discussion of anything like 4K TV. . .the list can go on and on.

Before considering the various freedoms that the audiences and the television industries will enjoy, it is important to first look at what television in 2025 and beyond will most probably look like. Each of the aspects listed is discussed in detail in the book *21st Century Television: The Players, The Viewers, The Money*.

- By 2025, the television networks will be delivering their programming directly to the consumer, divorcing themselves of their local affiliates as unnecessary baggage, and keeping whatever retransmission consent fees there are for themselves.

- Cable companies will become one of the major IP backbones for delivering all of television to viewers throughout the country.

- The major telephone companies such as AT&T and Verizon will be even more powerful competitors to the cable companies for IP backbone delivery.

- The continuing retransmission consent fee battles will have forced monthly cable, satellite, and telecom subscription prices to rise to the point that consumers will have finally rebelled and massive numbers of viewers will have moved to over-the-top options. As more and more viewers move to alternative methods of watching television, the content providers – primarily the cable channels at first, then the networks – will begin providing strong alternative opportunities for viewers to watch their programming over the Internet. The viewer rebellion will negatively affect both the content providers and the cable/satellite/telecom companies to such an extent that both groups will be forced into a much closer relationship.

- As such, at least a large portion, if not all, television will be delivered through some form of IPTV and will be delivered as video-on-demand.

- At the same time, the developing consolidation in the television industries that is in its early stages today will be rapidly moving to a time farther in the future where there will be only a handful of major integrated players left to compete against each other. While that time is much more distant (perhaps 2050 or 2070), ultimately, major holding companies consisting of television networks, cable companies, satellite companies, telephone companies, production houses, cloud providers, and other ancillary services will drive a global television industry that will reach viewers no matter where they are.

- More and more television programming, production, news, etc., will be in the cloud, opening new avenues to reach audiences around the world.

- Set-top boxes of all types will be in the process of being phased out as cloud delivery makes them unnecessary.

- Networks and cable channels will reside side-by-side on connected television sets, tablets, and smartphones (as well as Google-glass-type products, wearable technologies, etc.) as apps, widgets, or whatever is the current delivery mechanism of the day.

- Networks/cable channels will deliver their programming through websites – their own or through specialized sites designed specifically for their programming and will make full use of those apps, widgets, or current delivery mechanism on those connected TVs, tablets, smartphones, etc., to reach their audiences anywhere they might be.

- Much, if not all, of the advertising on television will be interactive, customizable, and targeted. Revenues will be approaching, if have not already reached, the tipping point where such advertising is preferable and more lucrative than traditional television advertising.

- Product placement on television will be well on its way to becoming ubiquitous with everything in every frame becoming a clickable link for the viewer.

- Search will be as easy as "1, 2, 3" – or "1, 2" – or even "1".

While not all of the aspects mentioned above may have come to complete fulfillment, the coming years leading to 2025 will be the leading edge of the revolution that is 21st Century Television. Looking at the list of aspects mentioned above, it is easy to see how Television In The Cloud will be the linchpin that binds those developments together.

So what are the freedoms that will be available in the near future? As mentioned earlier, those freedoms can be broken down by the groups that Television In The Cloud reaches – the viewers as well as the television industries themselves. Each of the groups will be discussed separately. First, though, a caveat: The freedoms are not possibilities that have never been discussed. The reader will likely recognize every one. However, thinking about the following as freedoms will perhaps put a new perspective and maybe even a heightened importance onto them for the reader.

Viewer Freedoms

Let's begin with the viewer - 21st Century Television driven by Television In The Cloud provides the viewer with a number of freedoms that are not possible in today's television viewing world. When one thinks of 21st Century Television, it's a world where the viewer can watch television anywhere, anytime, on any platform. Television In The Cloud makes it possible for those things to happen. So what are the freedoms the viewer can expect from this exciting new television universe?

Freedom of Personal Choice

Television In The Cloud provides the viewer with the freedom to watch his or her favorite program in the way in which (s)he wishes to watch it. The viewer is in total control of the television viewing decisions. (S)he has the ability to choose the preferred platform. If the viewer wishes to watch a program on a tablet, that's fine. If it's on a television set, that's fine as well. It's also the same with a mobile phone. If the viewer wants to watch the program at home, (s)he can. If the viewer wants to watch while traveling by car, train, or plane, that is also possible, if there is connectivity available. The viewer has the freedom of personal choice because the programming is available all the time, everywhere.

Today it is possible to see the start of this freedom, but it is in its infancy compared to what it will be in the coming years. Currently, the cable, satellite, and IPTV companies are offering to their subscribers the ability to view current television programming through the use of specialized apps (such as Comcast's "Xfinity" app, or DirecTV's app) for tablets and smartphones. Additionally, currently making its way through the U.S. federal court system is a case involving Aereo, a much more controversial innovation. Aereo uses micro-antennas to deliver live local television through the Internet to subscribers' tablets and smartphones within the viewing market of the local stations. Both the app offerings as well as Aereo are examples of first steps in this freedom that are being taken. Make no mistake - it is a freedom that every viewer desires and it will be a driving factor in the move to Television In The Cloud.

Freedom from Tyranny

Television In The Cloud provides the viewer the freedom from tyranny. Yes, that is correct, tyranny. There are actually several tyrannies that will be discussed in this section.

a. It's the freedom from the tyranny of appointment viewing. Television in the Cloud makes it possible for all programming to be delivered through video-on-demand, leaving it to the individual viewer to determine when he or she will watch a particular program. There is no longer any need to wait – for example – until 9:00 p.m. on a Wednesday night to watch an episode of the viewer's favorite program. With Television In The Cloud, the viewer watches the program when, where, and how (s)he wishes to watch. This form of viewing is already occurring in delayed form on Hulu/Hulu+ and various content provider websites such as CBS. ABC is experimenting with a form of on-demand viewing with the new WatchABC app, so there are early moves underway in this direction. What Television In The Cloud provides, though, is the opportunity to watch a program at any time without having to worry about if the viewing is delayed. The viewer can watch when the program or episode is released or can choose to wait until a later time of his or her choosing.

b. Going along with the freedom from the tyranny of appointment viewing is the freedom from the tyranny of television schedules. When all television programming is in the cloud and is available through video-on-demand, there is no longer a need for a programming schedule. All programs are available for the viewer, all the time. Of course, there will still be certain types of programs that will be scheduled and shown at the scheduled time – but they will be programs that are shown live, such as sporting events or certain reality programs. Otherwise, everything is delivered on demand for the viewer's convenience. New scripted programs will be listed by the content provider with a release date and time for those people who prefer to watch the show immediately upon its release. But regardless of whether it is a live program or a new scripted program, the viewer can choose whether or not to watch at the scheduled time or at any other later time. This freedom also exists for the television industries as well.

c. It's a freedom from the tyranny of having only one or two opportunities to watch a favored episode, and it's a freedom from the tyranny of having to wait until some programmer decides to air that favored episode. Traditional television has always limited the viewer to watching the program when the channel's programming department (network, local, or cable) wishes to air that program. Miss it and the viewer may have to wait weeks or months to watch the episode again, or maybe not – at least on the channel it originally aired, if aired on a broadcast channel – due to the death, for all intents and purposes, of the rerun on those channels. Even today, unless the viewer has purchased a service such as Netflix or Hulu+ or has an iTunes or Amazon Prime Video account, (s)he is likely stuck with waiting until there is that rare broadcast channel rerun, or the episode comes on a cable channel, where the viewer still has the same tyrannies listed above. Television In The Cloud eliminates that by not only allowing the viewer to watch when and where (s)he wants, but also *how many times* that viewer may want to watch and at the viewer's leisure. Today, the ability to watch a program episode more than once is generally only possible if (1) the viewer has cable, satellite, or telecom IPTV delivery to the home and one of the channels is running that episode, (2) the viewer purchases or rents the episode or the program season either online or at a retail outlet, or (3) owns a DVR with enough room left on it to record the episode. The first option is limited by the first two tyrannies. The second option requires additional expense. The third option is the next tyranny to be discussed. The freedom from the tyranny of having only one or two opportunities to watch a favored episode, and the freedom from the tyranny of having to wait until some programmer decides to air that favored episode puts the viewer in control of the viewing.

d. It's also a freedom from the tyranny of having to remember to set a DVR to record a program the viewer will miss because of another obligation, and it's a freedom from the tyranny of having to set a DVR for one program because there is a conflicting program on at the same time. How many times have you, as a viewer, had to remember to set a DVR because there was a conflict such as enjoying a dinner out, or attending a movie or a sporting event, or had some other conflict when the program came on? How many times have you forgotten to set a DVR when you were enjoying that dinner out, that movie or sporting event, or that something else when the program came on and so you missed it? A second scenario - How many times have you, as a viewer, missed a portion of, or all of, one program because you forgot to set the DVR when you were watching another program, couldn't record a program on your DVR because it was full, or there were several programs that you wanted to record but were limited in the number of programs that the DVR could record? For the Television In The Cloud viewer, there is no need for a DVR because all the programming is always available in the cloud. With Television In The Cloud, the DVR becomes an unnecessary piece of equipment that will be destined for the museums of technology.

e. Finally, it's a freedom from the tyranny of program packages that require a viewer to purchase 500 channels of programming when he or she really only watches or wants to watch 10. At this time, especially in the U.S., pay TV in the form of cable or satellite dominates the delivery of television. Eighty percent of television today is viewed either through cable or a DTH satellite system. When the IPTV systems are included, that number jumps to slightly more than 90%.[1] If viewers today want to watch more than whatever channels they can receive through over-the-air television (and 90+ percent do want more), they are forced into subscribing to the cable company that has the local monopoly where they live, to one of the two DTH satellite companies, or to one of the two telecom IPTV services *if* it is available where they live. Once the viewer has subscribed, (s)he has to then select from a list of packages where the viewer has to purchase a package of 100 or more channels to get the few (s)he wants to watch. Television In The Cloud eliminates those packages because the viewer is able to select any program (s)he wants from the cloud. Like the DVR, Television In The Cloud makes program packages obsolete and unnecessary, and television viewing for everyone becomes a de facto form of a la carte.

Freedom of Preferred Advertising

With Television In The Cloud, the viewer has the freedom to watch advertising that makes sense for his or her interests, wants, and needs, instead of having to endure advertising that has no relevance to, or does not connect with, the viewer. No longer is there any need for the broadcasters and cable channels to blast out commercials in hopes someone will respond to them. Through Television In The Cloud, the viewer receives only those commercials that are for things (s)he needs or wants. In reality, the freedom to watch customizable, addressable advertising is a win-win scenario for everyone.[2]

Freedom to Watch Everywhere

Finally, it's a freedom for the viewer to watch a program on the go, unencumbered by time, location, or lack of a television set. Television In The Cloud makes it possible for the viewer to enjoy television programming – his or her preferred programming – at any time of the day or night, weekday or weekend, one time or again and again, episode by episode, or an entire season or the entire series at one sitting, from anywhere in the world that the viewer can get a connection. Television In The Cloud also makes it possible to watch television programming on a tablet, a smartphone, a laptop computer, a desktop computer, on a television set, or on any other device available at the time the viewer chooses to watch television. Television In The Cloud makes it possible to watch in a hotel room in a different town, city, or country; in an Internet café in a rural area or on an island; while traveling domestically or internationally by private or commercial vehicle; all at the time and in the way that the viewer chooses. This freedom is, in reality, the culmination of all the other freedoms, because, ultimately, for the viewer, 21st Century Television is about the freedom to enjoy television – all of television – on his or her time schedule, when he or she wants to watch, where he or she wants to watch, on whatever platform is most convenient or preferred by the viewer.

Industry Freedoms

Now let's turn to the television industries. Television In The Cloud also provides a number of freedoms for the television industry.

Freedom from Schedules and Scheduling Strategies

Television In The Cloud provides the industry with the freedom from programming schedules and programming strategies. Traditionally, television has been built around the program schedule – a list of programs each broadcast network, local broadcast station, and each cable channel will air at specific times throughout the day, every day. Television programmers must plan their programming schedules using a variety of strategies to try to reach either the largest audience possible (the broadcast networks and local affiliates) or a specific target audience (the cable channels). Those strategies include such options as blunting, counterprogramming, lead-in, lead-out, hammocking, tent-poling, and numerous others.[3] The successful programmer was the one who could combine the correct program schedule with the correct program strategy that would allow him or her to maximize the audience desired by the advertisers for those programs.

With Television In The Cloud, the entire need for a day-long program schedule is removed. No longer does the content provider have to worry if it has applied the right scheduling strategy to the day's programming hours – no more concerns for lead-in, lead-out, hammocking, tent-poling or all those other strategies that are taught in programming classes at universities everywhere. With the use of Television In The Cloud, the whole inventory of a content provider is available to the audience at all times. All that's important for the content provider is to make sure the viewing audience knows what programs are available and how to get to them. As such, promotion and search become paramount for the programmer, so that viewers know what programs are included in the content provider's inventory, how they are cross-listed for search purposes, and – in the case of live programming such as live sporting events and live reality shows – what day and time they will air live for those viewers wishing to watch the live event in real time. Additionally, programmers will be tasked with the promotion and search responsibilities for new programs that are added to the content provider's inventory, and new episodes of current programs, so that the viewer can know the date and time of the initial airing of each new program or episode.

Freedom to Connect With the Viewer

Television In The Cloud provides the industry with the freedom to connect with the viewing audience through promotional techniques that allow each content provider to truly understand who the audience is for every program, right down to each individual in that audience. Television In The Cloud makes it possible to bring together all of the considerable forces of social media and social television to build in-depth profiles of the content provider's viewers and to reach out to the viewer in ways not possible before. From Twitter tweets, to Facebook "likes" and "fan pages," to LinkedIn connections, groups, industries, and companies followed, to other and newer forms of social media opportunities, even to YouTube and other such video sites, the opportunities for connecting with the viewer deeply and personally have never been greater. Television In The Cloud makes it possible for the content provider to have the ability to understand the viewer; to know the viewer's wants, needs, and desires; and to deliver to each viewer the *promotional information* (s)he needs to have the best and most enjoyable viewing experiences possible. By providing the ability to reach out to the viewing audience in ways that are deeply personal to each viewer, Television In The Cloud makes it possible for the content provider to have the freedom to build a lasting relationship with the viewer. In doing so, (s)he can experience a rich closeness to that content provider and believe that the provider has the viewer's best interests at heart.

Freedom to Effect New Viewing Patterns

Television In The Cloud provides the industry with the freedom to deliver programming that is right for each viewer. In traditional television, programmers provide a set schedule each day, consisting of programs that the ratings say the viewers want to watch. However, as discussed in detail in *21st Century Television: The Players, The Viewers, The Money*, the younger the viewer, the less likely he or she is to watch television on a TV set or at the appointed time.[4] More likely, those younger generations (generally about 45 years of age and younger!) watch television on their schedules using Hulu and Hulu+ for network programs; choosing Netflix, Amazon, or other online program providers for original content programs and movies; purchasing, renting, or downloading a season or series of a favorite program; or even going to user-generated programs on YouTube and other such sources.[5]

Television In The Cloud has the potential to make major changes in the current viewing patterns of young and old viewers alike. No longer does the viewer have to endure a night of nothing to watch on television, and nothing that's just right for him or her on

Netflix or other such program sources. The content provider's whole inventory of programs – from the newest programs to the oldest (and all future ones as well) – is available for the viewer to enjoy. The viewer can watch a single episode once, can watch a single episode again and again, or watch an entire series at one binge sitting if (s)he so chooses. With Television In The Cloud, the programs that viewers want to enjoy are there for them anytime, anywhere, using any platform, making it possible for the viewers to become even more loyal to those programs and to the content providers, because each program is right for each viewer.

Freedom to Increase Revenues and Profits

Television In The Cloud provides the industry with the freedom to make money – lots of money – even more money than the revenues of today – through advertising that works and works for everyone – the viewer, the content provider, and the advertiser. Traditional prime-time television advertising on the over-the-air networks has been, at its best, a shotgun approach designed to deliver a commercial for a product that generally has no more than about a 20% chance of being successful. The value of the ads is that 20% of a broadcast network's audience is still a large number of viewers.[6] For the cable channels, the percentages are somewhat higher, but the size of the viewing audience for their programs is significantly lower.

Television In The Cloud provides the industry with the ability to reverse those percentages for both the over-the-air networks and the cable channels. Television In The Cloud makes it possible for all content providers to deliver, through Aggregated Targeted Microadvertising, or ATMA, the personalized and customized advertising that the viewers want to watch, for products and services they care about, and only those products and services they care about. Traditional microadvertising for television is designed to target each individual viewer and deliver a specialized advertisement developed for that viewer and implemented to reach that viewer's preferences for products and services, using the personalities and creative techniques that the viewer most prefers. For television, the costs to try to reach each viewer individually are neither realistic nor cost-effective.

ATMA, on the other hand, makes it possible for content providers to accurately target groups of viewers with like attributes, delivering to each group advertisements for specific products and services preferred by those groups in the creative manner most appealing to each group. While still requiring multiple advertisements, ATMA – because of its group orientation – would require only a few, or at most, a handful of different advertisements for each product to reach a specified target audience instead of the millions of different advertisements that would be needed if using traditional microadvertising. The number of advertisements would depend, of course, on how broad or narrow the target audience might be – e.g., Coca-Cola would have a broad target audience full of different groups of viewers and so might need 6-12 different advertisements with different creatives. On the other hand, a narrowly targeted company – perhaps a company such as Kubota lawn equipment – might need only a couple of different ads to reach its entire target audience. With Television In The Cloud making it possible to stream different advertisements to different groups simultaneously, content providers could sell each commercial availability multiple times to different companies that had different products but whose group attributes intersect at the program.[7]

Television In The Cloud is also about the ability to deliver ubiquitous product placement, making it possible for the viewer to find out instantaneously about a product or service that (s)he sees on the television screen, whether it be an ad or something the viewer sees that is part of the program (s)he is watching. Product placement has been a part of the motion picture industry almost since its beginning, as is the case with product placement on television. Going back to 1950s television, high end luxury automobiles were a staple in certain detective programs, spy thrillers, and dramas involving the wealthy and those around them.

Today's product placement goes even further, with product placement opportunities appearing on certain reality programs (*American Idol* and *The Biggest Loser* are two well-known examples) leading into advertisements, to the point that it is often difficult to determine where the product placement/advertisement ends and the program begins. In fact, *The Biggest Loser* often seems more like a sixty-minute product placement than a true television program, given its constant references to some product brand used by the contestants on the program.

Television In The Cloud makes it possible to take product placement still further. Television programs delivered using the cloud can have every item in every frame of the program a clickable link. Using connected television sets, especially, viewers have the opportunity to "click" on any item on the screen and find out more about the product immediately, including what brand the product is (if necessary), information about the product and the brand, where to purchase the product, its costs, etc. For actors in the program, clicking on the face would let the viewer know more about who (s)he is, the actor's background, the actor's social media pages, etc. While Shazam today makes it possible for viewers to "Shazam" a program or advertisement for more information, product placement delivered by Television In The Cloud is ubiquitous and inherently more deeply informative and potentially more persuasive.[8]

Television In The Cloud provides the industry with the freedom to deliver advertising that has a 90% chance of success instead of the 10-20% chance of success of today's television through ATMA and ubiquitous product placement. It's a win-win-win situation for all concerned – the content provider, the advertiser, and the viewer.

Freedom from Boundaries

Finally, Television In The Cloud provides the television industry with the freedom from boundaries – local, regional, national, and even international boundaries. Television In The Cloud makes it possible for the television industry to reach their viewing audiences wherever they are, anywhere the world. No longer will international borders keep the television industry from their viewers except for the incidental spillover that occurs along the borders. All of television will have the same opportunity to deliver its programming to audiences that Netflix has today in reaching its subscribers in the countries in which it operates. For the content providers, television will truly become a global commodity, instantaneously available to everyone around the globe.

It's the freedom to deliver the programming and advertising that works for those now-global audiences. With little scheduling (live events) and complete video-on-demand, viewers, no matter where they may live or be visiting, will have the opportunity to enjoy their favorite programs. Further, the television industry will have the opportunity to develop new viewers throughout the world who are beyond the easy reach of traditional television. The ability to reach a world full of viewers opens new opportunities to develop new advertisers eager to reach those new viewers both in the home countries of the advertisers and around the world. For the content providers, the ability to develop new advertisers around the globe opens new revenue opportunities both at home and in countries throughout the world. The ability to develop global revenue opportunities, combined with Aggregated Targeted Microadvertising and ubiquitous product placement will magnify the inflow of revenue to the content providers to a point where the phrase "analog dollars to digital pennies" will become "analog dollars to digital trillions and more." As stated in the previous section, from a revenue standpoint – and now also from a global viewing standpoint – Television In The Cloud is a win-win-win situation for everyone.

Television In The Cloud truly provides the freedom to reach the whole world with programming and advertising, to meet the needs and desires of viewers in whatever city and country they may live or even be visiting. In doing so, Television In The Cloud gives the television industry the freedom to enhance the reputation as well as the bottom line of each content provider and the television industry as a whole throughout the entire world.[9]

Summary

Is Television In The Cloud something that's easy to accomplish? Obviously not, otherwise it would already be the delivery method of choice. There are still engineering details that need resolving. Legal problems such as digital rights management must be resolved. The question over how to enter or transcend the sovereign borders of each country of the world must also be resolved. Cultural questions of how to preserve each country's television industry(ies) must be answered. Privacy questions – the right of the individual to privacy versus the right of the businessperson to have the information necessary to reach his or her viewers with targeted advertising – must be balanced and determined.

Is Television In The Cloud worth accomplishing – absolutely. The benefits and opportunities discussed throughout this book offer numerous reasons for both the viewer and the television industries to embrace the development and implementation of Television In The Cloud.

Will Television In The Cloud usher in a new era – perhaps a new golden era of television – most likely. Television In The Cloud makes possible the ability to reach a global audience anywhere, anytime, on whatever platform is most convenient or preferred with programming that is right for the viewer. It can deliver advertising that is targeted and right for the viewer and product placement that makes it possible for the viewer to find out more about every object in the program (s)he is watching. It provides opportunities for the industry to develop promotional techniques that will allow the content providers and the viewer to connect in new and deeper ways. In short, Television In The Cloud makes the future of 21st Century Television limitless.

In his article "Embracing New Media Technologies," Jonathan Leess, Senior Vice president at Pac-12 Networks and former President of Digital Media at CBS Television as well as a former Senior Vice President at Disney/ABC, stated

> Throughout my career in broadcast television and digital media, I've tried to convey to my colleagues one thing: Content

will always be King but new emerging technologies must also be embraced. They are the critical components for delivering great content to a now, very different consumer; a consumer that is mobile, has more control and possess [sic] a strong power of influence....

Today, there's a critical shift taking place from the old mass reach and brand advertising approach of traditional television to targeting and engaging consumers, generating real-time data and establishing a direct relationship with the consumer....

With television still the most powerful medium in the world, content providers must recognize the new, empowered consumer; a consumer that is in control, wants to be connected to information and content that matters, that's relevant to their lives and they are highly influenced by the opinions of their peers.[10]

What Leess is expressing is the freedoms that the consumers of this new, developing age of television are expecting, and more and more are demanding. These freedoms can only be delivered in their fullness through the complete global implementation of Television In The Cloud.

To slightly modify the closing statement of the Epilogue of the book *21st Century Television: The Players, The Viewers, The Money*, "If there is a rallying cry for 21st Century Television, if there is a rallying cry for Television In The Cloud, and if there is a rallying cry for the exquisite future of this television universe that lies ahead of us, truly, that rallying cry must be "FREEDOM!!!!"[11]

CHAPTER 10
THE FUTURE OF TELEVISION IN THE CLOUD

Cloud technology is in a state of rapid development and improvement. It is a technology for both the present and the future. Television In The Cloud is in the relatively early stages of its development. As has been stated time and again in this book, television and the cloud are made for each other. To be sure, Television In The Cloud is being deployed today in increasing numbers by the legacy content providers, and is in use by the new media content providers.

However, much more must be done for Television In The Cloud to reach its full potential. Ultimately, Television In The Cloud will be the architecture upon which the entire television industry is built. That time, though, is still in the future. Much has to be done before Television In The Cloud becomes the basis for 21st Century Television.

Television In The Cloud As A Mindset

The major topic of discussion regarding the full deployment of Television In The Cloud is often the ability of the technology to deliver content all along the value chain from the development of the content to the viewing of the content by the consumer. However, the reality is that the engineering needed for full deployment of Television In The Cloud is already available and is in use by a number of different companies. Netflix, Hulu, Apple, and Amazon are just four of the most notable examples. Each of these companies has had the capability to deliver

HDTV-quality video to their subscribers. Netflix has announced that it will be streaming – during 2014 – UHDTV (or 4K TV) to its subscribers who own or will have purchased UHD television sets sometime during the year.[1] At CES 2014 – the Consumer Electronics Show in Las Vegas, NV – a variety of television set manufacturers demonstrated the UHD television sets that they will bring to market during the year for consumer purchase.[2]

Additionally, the tablet market has fully embraced the idea of Television In The Cloud. It's now possible for viewers to download and watch both feature films and television programs while on the go, using their favorite iPads, Samsung Galaxy Tabs, Amazon Kindle Fires, or any of a number of different tablets currently on the market. Smartphones continue to grow in screen size in part due to the desire of users to watch video – and, increasingly, television programs – on their phones as they travel. The continuing development of "phablets"[3] and the surrender of Apple to the fact that consumers want iPhones with significantly larger screens is testament to the power of the viewer's desire to watch television anywhere, anytime, and on the platform of his or her choosing. The underlying driving force is Television In The Cloud.

So the engineering basics of Television In The Cloud have been, for all intents and purposes, settled. Certainly, there will be the necessity of changes and improvements along the way, but, from an engineering standpoint, Television In The Cloud is a reality. What's now required of the television industry is a change in mindset from fear and doubt to embracing the new technology. This change in mindset must occur all along the value chain from the content producers,[4] to the content providers,[5] to the content distributors,[6] if the end user – the television viewer – is going to be satisfied with his or her 21st Century Television.

The Broadcasters

It is difficult for the legacy media[7] to change their mindsets from what has been their enduring way of producing, storing, and delivering television to their audiences. For the networks and their local affiliate stations, the current system has provided the industry with jobs, profits, and audiences going back to the late 1940s or early 1950s, especially following the end of the 1948-1952 television freeze.[8] For more than sixty years, the television networks have delivered programming to their local affiliates, which have then used their over-the-air signals to deliver the networks' programming into the homes of their viewers, to the cable systems in each market, or to the DTH satellite companies such as DirecTV and Dish Network for delivery to the viewers' homes in each market. These legacy media have worked together to make television a profitable industry for each for many years. It is understandable why these media view with skepticism the idea of changing their businesses from a way of producing, storing, and delivering television to their audiences that has worked for more than half a century to a system that is just coming into use today and has a highly anticipated, but largely untried future. For the broadcasters, especially, the thought of making the wholesale change to 21st Century Television delivered by Television In The Cloud is one to be viewed with severe trepidation. After all, the current system still delivers the largest audiences of any of the content deliverers today, despite the large number of viewing alternatives to the networks' programming. The mantra of "trading analog dollars for digital pennies"[9] is still very much in the forefront of the broadcasters' minds.

However, the comfortable system the legacy media has enjoyed is coming under constant pressure from both internal and external forces. From inside, the battle for retransmission consent fees is straining the synergistic relationships among all the parties. The over-the-air (OTA) broadcasters are constantly battling with the cable and DTH satellite companies every time a contract comes up for renewal. Since the Congress established the retransmission consent fee option in the early 1990s, viewers have had to endure wars of words and blackouts of channels as each side vies to negotiate a contract that is good for it, regardless of the effect on the other side and, especially, the viewer.[10]

Possibly the most public confrontation between an OTA broadcaster and a cable company occurred in August 2013. In that confrontation, there was an escalating battle between CBS, and its owned-and-operated local stations, and its cable channels against Time-Warner Cable over a new retransmission consent fee contract. The battle, played out in some of the top markets in the country, resulted in blackouts of CBS properties in those major markets for a number of days, and damaged the local TWC systems in those markets. While the hostilities were resolved in time for the 2013 college and professional football seasons, the damage to TWC, especially, was significant. That damage has almost certainly guaranteed other, likely devastating, battles over retransmission consent fee contracts in the future. This ongoing battle is causing tremendous riffs between the two sides, a lack of desire to compromise and work together for the good of both (as well as their viewers), and destroying the working relationship and good will the two sides may have had in the past.

A second, less discussed, aspect of the ongoing retransmission consent fee war is the battle between the networks and the owners of their local affiliates.[11] Under the current law, retransmission consent fees are paid to the local stations (or their group owners), *not* to the networks themselves. Given that

(*1*) the networks' combined audience is a shadow of its former self;

(*2*) the amount of money involved with retransmission consent contracts total in the billions of dollars; and

(3) the networks supply the bulk of the most-watched programming of their local affiliate stations,

the networks believe they deserve, at the very least, the bulk of the retransmission consent fee revenues. Naturally, the local stations and their group owners do not agree. While this battle, for the most part, is played out away from the viewing public, the battle is real and it is causing significant tensions between the two sides.

Externally, the forces arrayed as alternatives to the legacy media are numerous, and – taken together as a whole – constitute a potentially powerful alternate set of choices for the television viewer in both the near and longer terms. Not surprising, these challengers are the early vanguard of Television In The Cloud. The leader, of course, is Netflix, with more than 30 million subscribers as of the end of 2013.[12] With its continually developing powerhouse lineup of movies and original programs, Netflix demonstrates the ability of a cloud-based content provider to be successful in reaching what is now a set of multinational audiences. Netflix's success also demonstrates the future potential of such providers to deliver outstanding content to worldwide audiences using Television In The Cloud.

Two powerhouse alternatives to Netflix are Amazon's Instant Video delivery capability through its Prime subscription service, and the combination Hulu/Hulu+[13] content delivery service. Both of these major alternative offerings are serious competitors not only to Netflix, but also to the legacy content providers. Amazon's Instant Video delivery system began as a cloud-based movie delivery service to rival Netflix's original service. Now, Amazon has begun sinking substantial sums of money into acquiring original programming. In doing so, Amazon will rival both Netflix's original program offerings as well as competing directly with legacy media content. Additionally, in January 2014, the *Wall Street Journal* reported that Amazon was considering developing both an over-the-top set top box along the lines of a Roku box and a potential pay-TV service designed to compete directly with all the legacy media.[14] While Amazon has denied the development of the pay-TV service,[15] the *Wall Street Journal* has not pulled back from its position.

Hulu/Hulu+ is a very interesting cloud-based service because its origins lie with the broadcast networks as a way to give viewers the opportunity to watch television programs on a variety of platforms, but on a delayed basis. Hulu/Hulu+ has been on the market twice since its introduction, but both times the negotiations for the sale of the service fell through. Today, Hulu/Hulu+ is in the early stages of being a competitive alternative to Netflix and Amazon through the purchase of original television programming. However, in doing so, it also becomes a competitor to the legacy media as well. The final disposition of Hulu/Hulu+ is still to be decided, but the next months to couple of years should be an interesting time for the Television In The Cloud service.[16]

Possibly the most controversial of the alternative services, one that potentially could have an enormous impact on the entire current television universe and change dramatically television in the 21st Century, is a company called Aereo. Aereo, founded by Chet Kanojia and funded, in substantial part, by Barry Diller's IAC/InterActiveCorp, is a New York City-based technology company that operates a particular type of Television In The Cloud service. Aereo, using millions of dime-sized antennae, makes it possible for its subscribers to watch over-the-air local television on their tablets and smartphones using an Internet connection. The service assigns two antennae to each subscriber, one to allow the subscriber to watch the programming in real time and the second to record the program for later viewing, using the cloud to store the recorded programming. At this time, Aereo is available in 12 markets with plans to operate its service in 23 markets across the U.S.[17]

Currently, Aereo is engaged in litigation with just about all the broadcasters, both the local broadcast stations in every market where Aereo operates as well as the networks. At issue, once again, is the matter of whether or not Aereo's service constitutes a form of alternative television that requires Aereo to pay retransmission consent fees. Aereo argues it does not have to pay the fees; the broadcasters disagree.[18] At this time, the various cases involving Aereo and a direct competitor, FilmOn X, have reached the United States Supreme Court, which will consider the case during the current term, scheduling oral arguments for April, and issuing – what will be highly watched, second-guessed, and anticipated – a ruling by June 2014.[19] Depending on how the Court rules and the scope of the ruling, it is possible that the demise of the current state of television could be significantly hastened and Television In The Cloud could become the major architecture for 21st Century Television much sooner than anticipated.

The combination of internal strife and external forces weighing heavily on the various legacy media is destined to radically change the landscape of 21st Century Television. For the broadcasters, the combination is destined to tear the industry apart, sundering the networks from the local broadcasters and leading the networks into a new era. In this new era, the networks will deliver their programming directly to the viewers, through the use of cable and DTH satellite and various cloud-based options, some of which were discussed in Chapter 9. Ultimately, the broadcasters will find themselves moving completely to Television In The Cloud. This move will open new opportunities for the networks to reach audiences not possible before, and will allow them to develop new avenues for revenue that will lead them to greater levels of profitability and a new and exciting "golden era" of television.

The Cable Industry

For the cable industry, the mindset of the future is quite different. The cable industry has been much more open to the idea of the changing landscape of television, becoming the first to offer TV Everywhere scenarios to their subscribers. For the industry, more and more, the different cable companies are realizing that the future of cable is not cable but broadband. Cablevision, for one, is already considering moving to complete broadband in the near future.[20] HBO is considering offering its HBO Go as a stand-alone Television In The Cloud platform.[21] If HBO follows through on its consideration, in the future, viewers could subscribe to HBO Go without first having to subscribe to the cable system's premium HBO service. For cable companies, providing broadband service to their subscribers is the fastest growing part of the industry, even as the number and percentage of traditional cable subscribers continuously shrinks. The cable industry, more and more, is realizing the future, and is shifting from the traditional twisted-copper-wire cable mindset to a future where they, along with the IPTV services of AT&T and Verizon[22] will become the major content distributors of 21st Century Television through their ever-faster high speed broadband delivery.

The Content Producers

One of the more overlooked aspects of 21st Century Television, delivered by Television In The Cloud, is its effect on the content producers. During the early years of television (before 1975 and the advent of satellite delivery of cable programming), the three broadcast networks broadcast a maximum of 66 hours of prime time programming weekly (22 hours x 3 networks). Assuming that all programs were 30 minutes in length (which they were not), the maximum number of prime time programs during that time period would have been 132 episodes of programs.

Today, there are six broadcast networks[23] and approximately 800 national cable programming services,[24] all of which require programming throughout the day. For virtually all the cable services, there is a need for programming around the clock, requiring each service to provide 168 hours of programming each week. In addition, there are literally hundreds of niche channels providing programming to OTT devices such as Roku and other such devices. As the number of choices continues to increase, the need for programming from the content producers will continue to increase.

However, an equally overlooked aspect of 21st Century Television, driven by Television In The Cloud, is the fact that the program producers will have the opportunity – if they should choose to do so – to become their own content providers as well as producers. Television In The Cloud opens the way for the content producers of the future to use their own websites to provide programming directly to the viewer, bypassing the networks and cable channels completely.

Should the content producers decide also to become content providers – much as the broadcast networks do today – the decision would put the content producers in direct competition with their current partners. For the broadcast networks, which already produce programs, the change would have a limited effect on their programming decisions, especially in the medium to long term. They would simply produce more of their own programs. For the cable channels, however, especially those not connected with one of the broadcast networks or that regularly produce their own programming,[25] the effect would be significant, and, more likely, devastating. While the likelihood of the content producers also becoming content providers in the short term is virtually nonexistent, nonetheless, that option will likely weigh heavily on negotiations for program contracts in the medium to long term.

Also affecting contract negotiations, especially in the longer term, are the possibilities that Television In The Cloud opens for content providers to make their programs available to audiences around the world. In Chapter 9, one of the freedoms for the viewer is the ability to enjoy his or her programs anywhere, regardless of time, location, or lack of a television set. For the viewer, the term "location" refers to the ability to watch a favored program anywhere in the world where (s)he might be at the time. In other words, Television In The Cloud brings down the barriers limiting viewer enjoyment of watching that favorite program on the viewers' terms.

Chapter 9 also discusses a similar freedom for the television industry as well – the freedom to reach the viewer wherever (s)he is in the world and to open new opportunities to reach potentially *all* 21st Century Television viewers no matter where in the world they might be. Television In The Cloud opens the avenues for content providers (and, by extension, the content producers as well) to develop worldwide audiences for their programming far beyond the capabilities of today. The ability to reach these worldwide audiences at any time and/or simultaneously also opens new avenues to develop extremely lucrative opportunities for revenues of all types, leading the content providers to even larger worldwide profits than today.[26] For the content producers, the opportunities to significantly enhance their bottom lines by providing worldwide audiences with their programming directly, might just be enticing enough to make the content producers want to become content providers themselves.

So What's The Future?

In his September 4, 2011, article (remember, this is 2½ years ago!), Erick Shonfeld opens with these words: "TV is moving to the cloud. It is inevitable, just as other kinds of media from books to music are increasingly delivered over the Internet."[27] He finishes his article with this prediction:

> Eventually TV won't be the same unless it is online and connected to everything else. A show that can't be shared

or linked to will command less and less of our attention. Scheduled TV will go away for everything except live events like breaking news, sports, and award shows. [Author's note: Additionally, shows – primarily reality shows such as *Dancing With The Stars* and *American Idol* or the finale of shows such as *The Biggest Loser*, that require audience participation in some manner - will still be available to viewers in a live format as well.] The Internet will become our DVR, but one freed from the awful user interface of the current program guide. There is no grid large enough to contain all the video content you might want to watch, and why should your program grid be the same as mine? Trust me, it will be much better once it's all in the cloud.[28]

Television was made for the cloud, and the cloud was made for television. 21st Century Television demands a 21st Century solution to the delivery of its programs, advertising, promotional campaigns, etc. That solution is Television In The Cloud.

Television In The Cloud is the future of 21st Century Television. It will be the architecture of 21st Century Television and will open avenues for television to grow, change, improve, and enjoy a new golden age. For the viewer, Television In The Cloud will make it possible to enjoy television in whatever way (s)he chooses to enjoy it – anywhere, anytime, on whatever platform is most convenient or preferred. Further, the viewer will be able to enjoy viewing in every way, from the passive "lean-back" style of viewing of today, to a completely interactive, fully immersed "lean-forward" experience of a television universe that is only now in the earliest stages of development and experimentation. For the television industry, Television In The Cloud will provide a means to reach their audiences in ways not possible today. It will provide the means to increased profitability. It will make it possible for the industry to bring together the entire world to enjoy entertainment, share news, and touch lives in new and exquisite ways.

As the rampant changes in television technology continue to develop and grow, Television In The Cloud will play an ever-increasing, crucial role in the implementation of 21st Century Television. By 2020-

2025, expect television to move primarily to the cloud, and by 2025-2030, expect to see all of television as Television In The Cloud. After all, television *is* made for the cloud, and the cloud *is* made for television.

NOTES

Chapter 1 – Introduction to the Cloud

1. "Cloud Computing," *Oxford Dictionaries*, http://oxforddictionaries.com/definition/cloud+computing
2. "Cloud Computing," *Dictionary.com*, http://dictionary.reference.com/browse/cloud+computing
3. http://www.techterms.com/definition/cloud_computing
4. Poelker, Chris, "Defining cloud computing, part one: Laymen's terms," *Computerworld*, (April 6, 2012), http://blogs.computerworld.com/19959/defining_cloud_computing_part_one_laymen_s_terms
5. Mell, Peter and Timothy Grance, *The NIST Definition of Cloud Computing*, Special Publication 800-145, September 2011, http://csrc.nist.gov/publications/nistpubs/800-145/SP800-145.pdf
6. Hurwitz, Judith, Kaufman, Marcia, Halper, Fern, and Bloor, Robin. *Cloud Computing for Dummies*, Hoboken, New Jersey (John Wiley& Sons, Inc.), 2010, p. 9.
7. "Cloud Computing," *Reference.com*, http://www.reference.com/browse/cloud+computing
8. Some cloud experts argue that clouds are either public cloud or private clouds. They would argue that the other two are merely subsets of public and private clouds. Their view runs into problems when looking at hybrid clouds. Others include only hybrid clouds along with the public and private clouds. They argue that community clouds are subsets of one of the other clouds.
9. For more information, see Aycock, Frank A., *21st Century Television: The Players, The Viewers, The Money*, (Createspace/Amazon), 2012.
10. "Vannevar Bush," *The Electronic Labyrinth*, http://www2.iath.virginia.edu/elab/hfl0034.html
11. Bush, Vannevar, "As We May Think," *The Atlantic*, (July 1, 1945), http://www.theatlantic.com/magazine/archive/1945/07/as-we-may-think/303881/4/, retrieved July 11, 2013.
12. ibid.
13. "J.C.R. Licklider And The Universal Network," *Living Internet*, http://www.livinginternet.com/i/ii_licklider.htm
14. "The History of the Cloud," *The Official Livedrive Blog*, (August 30, 2012),

http://blog.livedrive.com/2012/08/the-history-of-the-cloud/

15. Mohamed, Arif, "A History of cloud computing," http://www.computerweekly.com/feature/A-history-of-cloud-computing

16. "The History of the Cloud," op. cit.

17. Biswas, Sourya, "A History of Cloud Computing," *Cloud Tweaks,* (February 9, 2011), http://www.cloudtweaks.com/2011/02/a-history-of-cloud-computing/

18. Dr. Chellapa is currently an associate professor and Caldwell Research Professor in the Information Systems & Operations Management area at the Goizueta Business School at Emory University in Atlanta, Georgia.

19. "Ranmath K. Chellapa," Emory University Directory, http://www.bus.emory.edu/ram/

20. Mohamed, op. cit.

21. "Amazon Elastic Compute Cloud," *Amazon Web Services,* http://aws.amazon.com/ec2/

22. Mohamed, op. cit.

23. Biswas, op. cit.

24. Anderson, Chris, FREE!: *The Future of a Radical Price*, New York (Hyperion), 2009.

25. Eucalyptus Corporation, http://www.eucalyptus.com/

26. "Cloud Computing: Past, Present, and Future," http://en.community.dell.com/cfs-filesystemfile.ashx/__key/communityserver-blogs-components-weblogfiles/00-00-00-00-11/4061.Cloud_5F00_computing_2D00_full.jpg

27. Biswas, op. cit.

28. Openstack, http://www.openstack.org/

29. Discussions regarding television in the cloud will be subject of Chapters 7 through 10.

30. The disagreements as to the development of television in the cloud can easily be seen simply by monitoring the different cloud-oriented groups on LinkedIn. They can become very intense and, at times, very heated!

31. For a look at what the future holds for television in the 21st Century, see *21st Century Television: The Players, The Viewers, The Money*, by Dr. Frank A. Aycock.

Chapter 2 – Types of Clouds

1. Mell, Peter and Timothy Grance, *The NIST Definition of Cloud Computing*, Special Publication 800-145, September 2011, http://csrc.nist.gov/publications/

NOTES

nistpubs/800-145/SP800-145.pdf

2. Subramanian, Krishnan, "Public Clouds," A white paper sponsored by Trend Micro, Inc.,

http://la.trendmicro.com/media/wp/public-clouds-whitepaper-en.pdf

3. ibid.

4. ibid.

5. ibid.

6. Williams, Charlie, "The Public Cloud is Not Always the Most Secure Option for Your Business," *2x blog*, August 27, 2013, http://www.2x.com/blog/2013/08/virtualization/public-cloud-security-082713/

7. There are a number of articles discussing the same topic. A quick Google search will produce numerous articles on security concerns of public clouds.

8. Williams, op. cit.

9. Williams, op. cit

10. Williams, op. cit

11. Williams, op. cit

12. Grimes, Roger A., "The 5 cloud risks you have to stop ignoring," *InfoWorld – Security Central,* March 19, 2013, http://www.infoworld.com/d/security/the-5-cloud-risks-you-have-stop-ignoring-214696.

13. ibid.

14. Asay, Matt, "Security Concern Not Slowing Public Cloud Adoption," November 15, 2013, http://readwrite.com/2013/11/15/security-concerns-not-slowing-public-cloud-adoption#awesm=~onXFqFU4rkKwMC.

15. Schultz, Beth, "Public cloud vs. private cloud: Why not both?" *Network World*, April 4, 2011, http://www.networkworld-digital.com/networkworld/20110404?folio=32#pg32

16. ibid.

17. ibid.

18. *PC Connection*, "Cloud Survey Results," http://www.pcconnection.com/IPA/PM/Brands/Cisco/PCCB2B/~/media/F6D6A531FB6943ACB374E8B06C8B8397.ashx?v=1

19. "What is a Private Cloud?" *Interoute*, http://www.interoute.com/cloud-article/what-private-cloud

20. "The Agility Gap In Today's Private Clouds," *CIO.com*, http://www.cio.com/white-paper/739166/The_Agility_Gap_in_Today_s_Private_Clouds

21. "Cloud Computing: Self-Run Private Cloud," *PCConnection.com*, http://www.pcconnection.com/IPA/PM/Info/Cloud-Computing/Self-Run-Private-Cloud.htm

22. "Cloud Computing: Managed Private Cloud," *PCConnection.com*, http://www.pcconnection.com/IPA/PM/Info/Cloud-Computing/Managed-Private-Cloud.htm

23. "Cloud Computing: Dedicated Private Cloud," *PCConnection.com*, http://www.pcconnection.com/IPA/PM/Info/Cloud-Computing/Dedicated-Private-Cloud.htm

24. Mell and Grance, op. cit.

25. Viswanathan, Priya, "The Hybrid Cloud – Is it the Best Solution for Cloud Computing?", *About.com Mobile Devices*, http://mobiledevices.about.com/od/additionalresources/a/The-Hybrid-Cloud-Is-It-The-Best-Solution-For-Cloud-Computing.htm

26. Williams, Charlie, "Hybrid Cloud Computing Overview and Benefits," *2X Cloud Computing Software Blog*, http://www.2x.com/blog/2013/09/virtualization/hybrid-cloud-computing-090313/

27. "What is a Hybrid Cloud?" *Interoute*, http://www.interoute.com/cloud-article/what-hybrid-cloud

28. Mell and Grance, op. cit.

29. Janssen, Cory, "What is a Community Cloud?", *techopedia*, http://www.techopedia.com/definition/26559/community-cloud

30. Butler, Brandon, "Are community cloud services the next hot thing?", *Network World*, March 1, 2012, http://www.networkworld.com/news/2012/030112-are-community-cloud-services-the-256869.html

31. Mell and Grance, op. cit.

32. Janssen, op. cit.

33. Samuels, Mark, "Community clouds: why they're a step too far for organisations," *CloudPro*, August 23, 2012, http://www.cloudpro.co.uk/cloud-essentials/hybrid-cloud/4415/community-clouds-why-theyre-step-too-far-organisations

34. ibid.

35. ibid.

Chapter 3 – Cloud Service Models

1. Factor, A. *Analyzing application service providers*. (Sun Microsystems, 2002).

2. *The Bessemer Cloudscape*, http://www.bvp.com/cloud

NOTES

3. Dropbox, "Plans," *Dropbox*, https://www.dropbox.com/pricing
4. *WebEx Meetings Plans: Free, Premium, Premium Plus and Enterprise*, http://www.webex.com/plans/meetings-plans.html
5. Welcome to ApacheTM Hadoop®! http://hadoop.apache.org/
6. *Platform as a Service - Comparison List*, http://www.paaslist.com/
7. Mosley, M., Brackett, M. H. & Data Management Association, The *DAMA guide to the data management body of knowledge* (DAMA-DMBOK guide), (2010), http://www.books24x7.com/marc.asp?bookid=40389
8. Bartlett, E. R., *Cable Television Handbook*, New York:McGraw-Hill, 2000.
9. Sandvine, *Global Internet Phenomena*, https://www.sandvine.com/trends/global-internet- phenomena/
10. Vogel, H. L., *Entertainment Industry Economics: A Guide for Financial Analysis*, Cambridge University Press, 2007.
11. Television Production in the US Market Research | IBISWorld, http://www.ibisworld.com/industry/default.aspx?indid=1246
12. Comcast - Channel Line Up, http://www.comcast.com/customers/clu/channellineup.ashx?clu=909#
13. Gadget Lab, "505,347,842 YouTube Channels and Everything Is On," *Wired.com*, http://www.wired.com/gadgetlab/2012/08/500-million-youtube-channels/all/
14. Press room – YouTube, http://www.youtube.com/yt/press/

Chapter 4 – Advantages of the Cloud

1. Software & Information Industry Association, *Guide to Cloud Computing for Policymakers*, (white paper) c. 2011, http://www.siia.com/index.php?option=com_content&view=article&id=799:cloud-computing-benefits-better-security&catid=163:public-policy-articles
2. ibid.
3. SunGard, *Security In The Cloud*, (white paper series), c. 2013, http://learn.sungardas.com/rs/sungardavailabilitysvcslp/images/cloud-services-security-in-the-cloud-CLD-WPS-057.pdf
4. Asay, Matt, "Security Concerns Not Slowing Public Cloud Adoption," *Readwrite*, November 15, 2013, http://readwrite.com/2013/11/15/security-concerns-not-slowing-public-cloud-adoption#awesm=~opJ9Rwh9iN0Rko
5. "What is a Hybrid Cloud?" *Interoute*, http://www.interoute.com/cloud-article/

what-hybrid-cloud

6. Haff, Gordon, "Maximize Strategic Flexibility By Building An Open Hybrid Cloud," *RedHat*, March 20, 2013, http://www.redhat.com/rhecm/rest-rhecm/jcr/repository/collaboration/jcr:system/jcr:versionStorage/316faea70a05260137ee5cab232304bc/2/jcr:frozenNode/rh:pdfFile.pdf

7. Cormier, Paul, "Why Should Your Cloud Be Designed With Open And Hybrid in Mind?" *RedHat*, July 30, 2013, http://www.redhat.com/rhecm/rest-rhecm/jcr/repository/collaboration/jcr:system/jcr:versionStorage/d795d1170a0526011fdcc6bd5fb826e3/2/jcr:frozenNode/rh:pdfFile.pdf

8. Ross, Asher, "How Cloud Computing Benefits Your Company," *Smart Data Collective*, October 7, 2013, http://smartdatacollective.com/asher-ross/150686/10-ways-know-how-cloud-computing-benefits-your-company

9. Software & Information Industry Association, op. cit.

10. Jennings, Richi, "5 Benefits of Moving to the Cloud," *Webroot*, http://www.webroot.com/us/en/business/resources/articles/cloud-computing/five-financial-benefits-of-moving-to-the-cloud

11. Olavsrud, Thor, "How Cloud Computing Helps Cut Costs, Boost Profits," *CIO*, March 12, 2013, http://www.cio.com/article/730036/How_Cloud_Computing_Helps_Cut_Costs_Boost_Profits

12. OFlaherty, Kate, "Costs and benefits of moving to the cloud," *Techradar*, March 21, 2013, http://www.techradar.com/us/news/internet/costs-and-benefits-of-moving-to-the-cloud-1139250

13. For further information, see Chapter 2 – Types of Clouds.

14. Blaisdell, Rick, "Cloud computing enables business scalability and flexibility," *CloudTweaks*, September 6, 2012, http://www.cloudtweaks.com/2012/09/cloud-computing-enables-business-scalability-and-flexibility/

15. Arno, Christian, "The Advantages of Using Cloud Computing," *Cloud Computing Journal*, (April 14, 2012), http://cloudcomputing.sys-con.com/node/1792026

16. Kumar, Arun, "The Advantages of Cloud Computing," *Bright Hub*! (May 19, 2011), http://www.brighthub.com/environment/green-computing/articles/10026.aspx

17. Arno, op. cit.

18. Ross, op. cit.

19. Ross, op. cit.
20. Jennings, op. cit.
21. "Why Move to the Cloud? 10 Benefits of Cloud Computing," *Salesforce*, http://www.salesforce.com/uk/socialsuccess/cloud-computing/why-move-to-cloud-10-benefits-cloud-computing.jsp

Chapter 5 – Building Block of Television In The Cloud

1. Erl, T. *Service-oriented architecture: concepts, technology, and design*. (Prentice Hall Professional Technical Reference, 2005).
2. Erl, T., Mahmood, Z. & Puttini, R. *Cloud computing concepts, technology & architecture*. (ServiceTech Press, 2013).
3. Harold, E. R. & Means, W. S. *XML in a nutshell*. (O'Reilly, 2004).
4. Rovi Cloud Services - Rovi Cloud Services API Documentation. at <http://developer.rovicorp.com/docs>
5. Austerberry, D. *Digital Asset Management*. (Taylor & Francis, 2012).
6. Video Space Calculator - Digital Rebellion. at <http://www.digitalrebellion.com/webapps/video_calc.html>
7. AWS | Amazon Elastic Transcoder - Video Transcoding in the Cloud. at <http://aws.amazon.com/elastictranscoder/>
8. CableLabs Specifications Online Content Access Authentication and Authorization Interface 1.0 Specification (CL-SP-AUTH1.0-I02-110324) - CL-SP-AUTH1.0-I02-110324.pdf. at <http://www.cablelabs.com/wp-content/uploads/specdocs/CL-SP-AUTH1.0-I02-110324.pdf>
9. OATC > Home. at <http://www.oatc.us/>
10. Jezierski, A. *Television Everywhere: How Hollywood Can Take Back the Internet and Turn Digital Dimes Into Dollars*. (i2 Partners LLC, 2010).
11. Avails - Avails_v1.0.pdf. at <http://www.movielabs.com/md/avails/v1.0/Avails_v1.0.pdf>
12. Lee, J. *Scalable Continuous Media Streaming Systems: Architecture, Design, Analysis and Implementation*. (John Wiley & Sons, 2005).
13. Harte, L. *Advanced TV advertising*. (Athos Pub., 2011).
14. Rosenblatt, B., Trippe, W. & Mooney, S. *Digital rights management: business and technology*. (M & T Books, 2002).
15. UltraViolet FAQ. at <http://uvdemystified.com/uvfaq.html>

Chapter 6 – Moving Television to the Cloud

1. Orlebar, J. *The Television Handbook*. (Routledge, 2011).
2. Fiske, J. *Television Culture*. (Taylor & Francis, 2010).
3. Burns, R. W. *Television: An International History of the Formative Years*. (IET, 1998).
4. Chan-Olmsted, S. M. *Competitive strategy for media firms: strategic and brand management in changing media markets*. (Lawrence Erlbaum Associates, 2006).
5. Tozer, E. P. J. *Broadcast Engineer's Reference Book*. (Taylor & Francis, 2004).
6. Waggoner, B. *Compression for Great Video and Audio: Master Tips and Common Sense*. (Taylor & Francis, 2013).
7. Ghanbari, M. *Standard Codecs: Image Compression to Advanced Video Coding*. (IET, 2003).
8. Sarmiento, A. S. & Lopez, E. M. *Multimedia Services and Streaming for Mobile Devices: Challenges and Innovation*. (Idea Group Inc (IGI), 2011).
9. HTTP Live Streaming Resources - Apple Developer. at <https://developer.apple.com/streaming/>
10. Overview of MPEG-DASH Standard « For Promotion of MPEG-DASH. at <http://dashif.org/mpeg-dash/>
11. Held, G. *A Practical Guide to Content Delivery Networks, Second Edition*. (CRC Press, 2010).
12. *Embedded Software*. (Newnes, 2008).
13. Ciciora, W. S. *Modern cable television technology: video, voice and data communications*. (Elsevier/Morgan Kaufmann Publishers, 2004).
14. Hua, X.-S., Mei, T. & Hanjalic, A. *Online Multimedia Advertising: Techniques and Technologies*. (IGI Global, 2011).
15. Microsoft Word - VAST 2_0 Clean FINAL.doc - VAST-2_0-FINAL.pdf. at <http://www.iab.net/media/file/VAST-2_0-FINAL.pdf>
16. Vogel, H. L. *Entertainment Industry Economics: A Guide for Financial Analysis*. (Cambridge University Press, 2007).
17. *Connecting the Dots Between Consumers, Content and Consumer Electronics in the Home*. (Consumer Electronics Association, 2011).
18. The emerging trend in entertainment: Crowdfunding. *CrowdFund Beat* at <http://crowdfundbeat.com/the-emerging-trend-in-entertainment-crowdfunding/>

NOTES

19. Netflix stock soars on huge subscriber growth - Jan. 22, 2014. at <http://money.cnn.com/2014/01/22/technology/netflix-earnings/>
20. COMCAST CORP - comcast10K.pdf. at <http://files.shareholder.com/downloads/CMCSA/2935775837x0x650076/e95fd726-8a42-4ca9-afb3-dfbd95113b40/comcast10K.pdf>
21. *Media Trends*. (SNL Kagan, 2011).
22. Media, C. D. B. E. -mai. A. G. R. feed D., 19, T. P. S. & 2011. *Your Guide to Who Measures What in the Online Space*. at <http://adage.com/article/media/guide-measures-online-space/229858/>

Chapter 7 – The Cloud and Television

1. Aycock, Frank A. (2012), *21st Century Television: The Players, The Viewers, The Money, Charleston*, S.C. (CreateSpace), 135.
2. For more information on search and promotion, see "Chapter 18 – Promotion" in Aycock, op. cit., 279-289.
3. For those readers interested in finding out more about where the television industries are heading and what the future of television is, please see Aycock (2012), *21st Century Television: the Players, The Viewers, The Money*, Charleston, S.C. (CreateSpace).
4. For more information on advertising and ATMA, see "Chapter 15 – Advertising" in Aycock, op. cit., 241-257.
5. For more information on global possibilities of 21st Century Television, see "Chapter 19 – "Going Global TV: TV Everywhere Fulfilled" in Aycock, op. cit., 291-303.
6. Internet Protocol Television.
7. http://money.cnn.com/news/newsfeeds/gigaom/media/ 2010_08_23_the_future_of_tv_is_not_on_cable.html.
8. For more information on video-on-demand, see "Chapter 5 – Video-On-Demand" in Aycock, op. cit., 69-77.
9. For more information on set-top boxes, see "Chapter 6 – Over-The-Top Set Top Boxes" in Aycock, op. cit., 79-94.
10. For more information, see the section on Mobile DTV, in "Chapter 7 – Mobile DTV, Connected TV, And The iWorld" in Aycock, op. cit., 95-103.
11. For more information, see the section on Connected TV, in "Chapter 7 – Mobile DTV, Connected TV, And The iWorld" in Aycock, op. cit., 103-110.

Chapter 8 – Making Use of Television In The Cloud

1. Austerberry, David, "Is video editing in the cloud for real?" *Broadcast Engineering Blog*, January 2, 2013, http://broadcastengineering.com/blog/video-editing-cloud-real.
2. Giardina, Carolyn and Adrian Pennington, "Cloud-Based Tech No Longer Just Blue Sky," *TV NewsCheck*, October 11, 2012, http://www.tvnewscheck.com/article/62787/cloudbased-tech-no-longer-just-blue-sky.
3. Ochiva, Dan, "Does make.tv Really Show the Future of Cloud-based Production?" *NYC Production & Post News*, December 18, 2013, http://nycppnews.com/2013/05/make-tv-show-future-cloud-based-production/.
4. Dodson, Andrew, "News Services Moving From Satellite To Cloud," *TVNewsCheck*, October 3, 2013, http://www.tvnewscheck.com/article/70939/news-services-moving-from-satellite-to-cloud.
5. ibid.
6. ibid.
7. ibid.
8. ibid.
9. ibid.
10. For more information on ATMA, see "Chapter 15 – Advertising," in Aycock, Frank A. (2012) *21st Century Television: The Players, The Viewers, The Money*, Charleston, S.C. (CreateSpace), 241-257.
11. ibid.
12. For more information on ubiquitous product placement, see Chapter 16 – Product Placement, in Aycock, Frank A. (2012) *21st Century Television: The Players, The Viewers, The Money*, Charleston, S.C. (CreateSpace), 259-268.
13. For more information on promotional techniques, see "Chapter 18 – Promotion," in Aycock, op. cit., 279-289.
14. For more information on ubiquitous product placement, see "Chapter 19 – Going Global TV: TV Everywhere Fulfilled," in Aycock, op. cit., 291-303.

Chapter 9 – The Freedoms of Television In The Cloud

1. A. C. Nielsen, Co., *A Look Across America: The Cross-Platform Report*, December 2013, http://www.nielsen.com/content/dam/corporate/us/en/reports-

NOTES

downloads/2013%20Reports/The-Cross-Platform-Report-A-Look-Across-Media-3Q2013.pdf

2. For a much more in-depth explanation of how this freedom becomes available, see Aycock, Frank A. (2012), *21st Century Television: The Players, The Viewers, The Money*, Charleston, S.C. (CreateSpace), 241-257.

3. For a discussion of programming strategies see any number of Broadcast (or Electronic Media) Programming textbooks. One such is Eastman, Susan Tyler and Ferguson, Douglas A., *Media Programming: Strategies and Practices*, 8th ed., Boston (Wadsworth Cengage Learning), c. 2009.

4. For a much more in-depth explanation of viewing patterns see the section on "The Viewers" in Aycock, op. cit., 141-240.

5. ibid.

6. Aycock, op. cit., 241-257.

7. For a more in-depth discussion of Aggregated Targeted Microadvertising, see Aycock, op. cit., 245-257.

8. For a more in-depth discussion of ubiquitous product placement, see Chapter 16 – Product Placement in Aycock, op.cit.

9. For more information on the global aspect of 21st Century Television, see Aycock, op.cit, 291-303.

10. Leess, Jonathan, "Embracing New Media Technologies," *rthk mediadigest*, June 14, 2011, http://rthk.hk/mediadigest/20110614_76_122758.html.

11. Aycock, op. cit., 342.

Chapter 10 – The Future of Television In The Cloud

1. Nakashima, Ryan, "Netflix app to stream 4K on new TVs immediately," *NBC News*, January 10, 2014, http://www.nbcnews.com/technology/netflix-app-stream-4k-new-tvs-immediately-2D11897506

2. Warman, Matt, "CES 2014: wearables, UHDTV and tablets galore," *The Telegraph*, January 27, 2014, http://www.telegraph.co.uk/technology/ces/10549509/CES-2014-wearables-UHDTV-and-tablets-galore.html

3. The term "phablet" is a combination of the words "PHone" and "tABLET," and is used to describe a classification of smartphones with large screens, generally in the 6" diameter range. The best known of these phones is the Samsung Galaxy Note series, and is the forerunner of a class of future smartphones that will make the

need for both a smartphone and a tablet computer obsolete.

4. Content producers is the 21st Century Television term to describe both the traditional production houses, film studios, the television networks, and independent television producers, as well as the new producers of video for OTT and UGC platforms.

5. Content providers is the 21st Century Television term for the television networks, cable channels, OTT companies such as Netflix, Amazon, Hulu, as well as others, and any other company that provides video content to the viewer.

6. Content distributors is the 21st Century Television term for the broadcast stations, cable systems, DTH satellite companies, AT&T and Verizon's U-verse and FiOS, respectively, ISPs, etc.

7. The "legacy media" term includes the traditional over-the-air (OTA) broadcasters, the cable companies, and the direct-to-home (DTH) satellite companies.

8. Head, Sydney W. and Sterling, Christopher H, *Broadcasting in America: A Survey of Electronic Media*, 6th ed., Boston (Houghton Mifflin), 1990, 65-68.

9. Zucker, Jeff, "Keynote Address," 2008 National Association of Television Program Executives convention, January 28-31, 2008, Las Vegas, NV.

10. For more information on retransmission consent fees and the history of some of the contract wars, see "Chapter 17 –Retransmission Consent Fees" in Aycock, Frank A., *21st Century Television: The Players, The Viewers, The Money*, Charleston, SC (CreateSpace), 2012.

11. ibid.

12. Carter, Bill, "Strong Finish to 2013 for Netflix as Profits and Subscriptions Soar," *New York Times*, January 22, 2014, http://www.nytimes.com/2014/01/23/business/media/growth-of-netflix-subscribers-surpasses-analysts-expectations.html?_r=0.

13. www.hulu.com

14. Sharma, Amol, Ramachandran, Shalini, & Clark, Don, "Amazon Considering Online Pay-TV Service," *The Wall Street Journal*, January 21, 2014, http://online.wsj.com/news/articles/SB10001424052702304757004579334981130200324.

15. Barr, Alistair, "Amazon says it is not planning pay-TV service," *USA Today*, January 21, 2014, http://www.usatoday.com/story/tech/2014/01/21/amazon-pay-tv-service/4726757/.

16. www.hulu.com

17. www.aereo.com

18. Flint, Joe & Faughnder, Ryan, "Supreme Court to hear Aereo case," *Los Angeles Times*, January 11, 2014, http://www.latimes.com/entertainment/envelope/cotown/la-et-ct-aereo-supreme-court-20140111,0,5028844.story#axzz2rc4xrgvw.
19. ibid.
20. Ramachandran, Shalini & Peers, Martin, "Future of Cable Might Not Include TV," *Wall Street Journal* (Online Edition), August 4, 2013, http://online.wsj.com/news/articles/SB10001424127887323420604578647961424594702.
21. See Lieberman, David, "HBO GO Without Cable? Not Yet, Says Time Warner Chief," *Deadline New York*, May 1, 2013, http://www.deadline.com/2013/05/hbo-go-without-cable-not-yet-says-time-warner-chief/ and Harris, Aisha, "We're One Step Closer to Stand-Alone HBO Go Service," *Slate*, October 25, 2013, http://www.slate.com/blogs/browbeat/2013/10/25/hbo_now_available_without_premium_cable_subscription_through_comcast.html
22. U-verse is the IPTV offering for AT&T; for Verizon, it's FiOS.
23. The six broadcast networks today are ABC, CBS, NBC, CW, Fox, and Univision.
24. "Industry Data," *National Cable Television Association* (NCTA), http://www.ncta.com/industry-data.
25. There are cable channels that do produce their own original programming. One such is ESPN, which produces most of the content they telecast. Interestingly, the ESPN family of cable channels is also owned by one of the networks – ABC/Disney.
26. For a more detailed look at the possibilities for global opportunities, see "Chapter 19 – Going Global TV: TV Everywhere Fulfilled" in Aycock, op. cit., 291-303.
27. Schonfeld, Erick, "TV In The Cloud," *TechCrunch*, September 4, 2011, http://techcrunch.com/2011/09/04/tv-cloud/.
28. ibid.

INDEX

A&A see Authentication and Authorization
ABC, 46, 73
    abc.go.com, 46
    ABC NewsOne, 105
account management, 76
adaptive bitrate streaming, 78, 79
ad server, 70
Adobe, 104
Adobe Flash, 70, 71
Adobe FlashAccess, 70
ad-supported, 73, 85
advertiser, 85 - 87
advertising, 85 -87
advertising insertion, 76, 80, 81
Aereo, 90, 136
    Barry Diller, 136
    Chet Kanojia, 136
    IAC/InterActiveCorp, 136
aggregated targeted microadvertising, 91, 106, 107, 125, 126, 127
Akamai, 79
Allaire, Jeremy, 21
Amazon, 18, 21, 22, 42, 58, 68, 131, 135
    Amazon Instant Video, 135
    AWS service, 18, 21, 42
    Elastic compute cloud (EC2), 21
    Kindle Fire, 132
American Idol, 126
"analog dollars to digital pennies," 128, 133
"analog dollars to digital trillions and more," 128
analog to digital changeover in U.S., 115
analytics, 72, 76
Anderson, Chris, 21
    Free!, 22
antenna, 73
Apple, 70, 131
    iPad, 132
application service provider, 40, 45
archives, 63
ASP see application service provider
Athenahealth, 40
AT&T, 137
audience trends, 81, 82
authentication, 63
authentication and Authorization, 63, 67
authorization, 63
automation, 66, 67, 75
avails, 68
Avid, 104
Bessemer Venture Partners, 40
BestBuy, 39
BI see business intelligence
big data, 42
Biggest Loser, 9,16
billing account management, 126, 127
Bitcentral, 105
    Oasis system, 105
Brightcove, 21, 76
broadcast operations, 21, 76
broadcast systems
browsing and discovery, 68, 79, 80
buffer see video buffer

INDEX

Bush, Vannevar, 20
    Memex, 20
business intelligence, 72
business models, 73, 74
cable television, 73, 83
Cablevision, 137
CableLabs, 67
Capgemini, 30
catalog, 39
CBS, 73
    CBS Newspath, 104
CDN see content distribution network
CES 2014, see Consumer Electronic Show
changes in television by 2025, 115 - 117
Chellapa, Dr. Ramnath, 21
chunk, 78, 79
chunking, 78
CinemaNow, 68
Cisco/PC Connection, 30, 31
Cisco WebEx, 40
cloud advantages
    cost, 55 - 57
        benefits, 55, 56
        economic impact, 57
    flexibility, 52, 53
    green computing, 57, 58
        environmental benefits, 57, 58
    mobility, 53, 54
        use by teams, 54
    portability, 54
        importance in early days of WWW, 54
    scalability, 57
        definition, 57
    security, 49 - 52
        key practices, 51, 52
cloud computing
    aspects, 17, 18, 26
    definition, 15, 16
    economic impact, 58
Cloud Computing Conference – West, 113
Cloud Foundry, 43
CloudBees, 43
CNN Newsource, 105
Comcast, 47, 82
community cloud, 36, 37
    considered a fad, 37
    description, 36
    usefulness, 37
compliance, 72
compression, 77
Consumer Electronic Show, 132
content distribution network, 79
content libraries, 63 - 64
content owners, 82
content windows, 74, 75
    premium window, 74
Coyle, Joe, 30
crowd-funding, 82
culture, 83
DaaS see Data as a Service
DAM see digital asset management
Dancing With The Stars, 140
DASH see Dynamic Adaptive Streaming over HTTP
dashboards, 76
Data as a Service, 44 - 46

data-type model, 44
Database as a Service, 42
digital asset management, 64
digital rights management, 71, 76
digital store check, 72
digital supply chain, 75, 76
digital video recorder see DVR
discovery, 46, 47, 68, 69, 79, 80
distribution, 80, 84
DRM see digital rights management
Dropbox, 40, 41
DSC see digital store check
DTH satellite 133, 137
 DirecTV, 133
 Dish Network, 133
DVD, 68, 74
DVR, 73
Dynamic Adaptive Streaming over HTTP, 70, 78
EMA see Entertainment Merchants Association
encoding, 65
Engine Yard, 43
Entertainment Merchants Association, 68
Eucalyptus platform, 22
Evertz, 64, 66
eXtensible Markup Language see XML
file based workflow, 66, 75
FilmOn X, 136
Force.com, 43
formats, 65
Fox, 73
 FoxNow, 67
freemium, 41

Generation Technologies, 105
global delivery, 109 - 111
Gmail, 40
Google, 22, 42, 64
 Google App Engine, 43
 Google Apps, 40
 Google docs, 21, 22, 27
Grimes, Roger, 29
 ownership concerns, 29
h.264, 77
Hadoop, 42
Harris, 66
HBO, 137
 HBO Go, 46, 67, 1378
high-definition, 77
HITS see Hollywood IT Society
HLS see HTTP Live Streaming
Hollywood IT Society, 75
horizontal scaling, 42
HTML, 62, 70
HTTP Live Streaming, 78
Hulu/Hulu+, 68, 87, 88, 94, 95, 131, 135
hybrid cloud, 34 - 36
 description, 34
 features, 35 - 36
Hyper Text Markup Language see HTML
IAB see Interactive Advertising Bureau
IaaS see Infrastructure as a Service
IBM, 42
industry freedoms, 122 - 128
 freedom from boundaries, 127, 128
 freedom from schedules and strategies, 122, 123

INDEX

freedom to connect with the viewer, 123, 124
freedom to effect new viewing patterns, 124, 125
freedom to increase revenues and profits, 125 - 127
Infrastructure as a Service, 41 -42
ingest, 77, 78
Interactive Advertising Bureau, 81
internal cloud, 73, 74
international, 79
intranet, 19
    compare with Internet, 19
Info-Tech, 3
IPTV, 29, 88, 89, 92, 93, 110, 137
Jennings, Richi, 55
"lean-back viewing," 140
"lean-forward viewing," 140
Joyent, 42
Kaltura, 76
Leess, Jonathan, 129 - 130
legacy media, 23
    television, 23
Lewis, Margaret, 21
licensing, 68
Lickliter, J.C.R., 20
Limelight, 79
linear playback, 66, 67, 70
linear programming, 73
Linux, 43
low water mark, 80
Make.tv, 104
MAM see media asset management
manifest file, 80
Master, 64

McCarthy, John, 21
McLuhan, Marshall, 45
media asset management, 64
metadata, 68, 79
metadata management, 76
mezzanine file, 77
Microsoft, 22, 43
    Azure, 18, 22, 42, 43
    Microsoft PlayReady, 71
    Microsoft Silverlight, 71
mobile, 67, 83
monetize, 83
monetization, 83 - 85
MPEG-DASH see Dynamic Adaptive Streaming over HTTP
MPEG2, 77
National Institute for Standards and Technology, 16, 25
NBC, 73
Netflix, 45, 58, 68, 69, 82, 87, 88, 94, 95, 110, 128, 131, 134, 135
network conditions, 7
network operations, 72
News, 104, 105
    field news production, 104, 105
NewsOne, 105
Nielsen, A. C., 84
OATC see Open Authentication Technology Committee
Ooyala, 76
Open Authentication Technology Committee, 67
Open Shift, 43
operational support services, 72
OSS see operational support services

PaaS see Platform as a Service
paaslist.com, 43
Pandora radio, 96
per-use, 41
per-user, 41
Peters, Tom, 44
phablets, 132
piracy, 71
Platform as a Service, 43, 44
post-production, 64
ProRes 422, 77
product placement, 107, 126, 127
production (television), 102 - 104
    field production, 102, 103
    live production, 103
    studio production, 103
programming (television), 99 - 122
    storage, 100
    inventory access, 100, 101
    ease of viewing, 101, 102
promotion, 107 - 109
    search, 108
    social media, 108 - 109
    other online sites, 109
    website promotion, 108
private cloud, 30 - 34
    benefits, 31
    dedicated, 33, 34
    managed, 32, 33
    self-run (standalone), 32
public cloud, 26 - 30
quality control, 71, 72
Rackspace, 18, 22, 42
    Openstack, 22
rallying cry, 130

reassembly, buffering, and rendering, 80
recommendations, 69
resolution, 77
retransmission consent fees, 133, 134
Roku, 135, 138
SaaS see Software as a Service
sales/advertising, 106 - 107
Salesforce.com, 21, 40, 43
Samsung Galaxy Tab, 132
search, 64, 76
service oriented architecture, 61, 75
set-top box, 46
Shazam, 127
Shonfeld, Erick, 139
Sirius/XM, 96
SOA see service oriented architecture
SocialCam, 88
Software as a Service, 39 - 41, 46, 47
STB see set-top box
standard-definition, 77
streaming, 69, 70, 76, 77
streaming video, 76
streaming video server, 76
sub-master, 64
Subramanian, Krishnan, 27
    business agility, 28
    cost savings, 27
subscription video-on-demand, 75
super-high definition television, 23
sVOD see subscription video-on-demand
targeting
technical support, 72
TED talk, 113
TEDx convention, Nagoya, Japan, 113

INDEX

Television as a Service, 45 - 47
Television In The Cloud
    as a mindset, 131 - 139
    broadcasters, 132 - 137
    cable industry 137
    content producers, 137 -139
    controversy, 114
    impact on television industry, 89 - 96
    legacy industries, 90 - 93
    broadcast networks, 90, 91
    cable industry, 91, 92
    telecom industry, 92, 93
    global, 89, 92
    new media industries, 93 - 96
    mobile DTV, 96
    set top boxes, 95
thePlatform, 76
Time-Warner Cable, 134
Tribune, 44
trafficking, 66, 67, 75
transactional, 73, 76
transcoding, 65, 66, 77, 78
transfer, 65
TV Everywhere, 46, 67, 137
TV Guide, 44, 69
TVaaS see Television as a Service
ubiquitous product placement, see product placement
ultra-high definition television, 23, 77, 89, 132
Ultraviolet, 71
VAST see Video Ad Serving Template
V2Solutions, 44
Verizon, 137

vertical scaling, 42
VHS, 73, 74
Video Ad Serving Template, 81
video buffer, 80
VideoFusion, 105
video management platform, 76
video-on-demand, 88, 93
    change to television-on-demand, 94, 95
video platform see video management platform
viewer addressable advertising, 106, 107
viewer freedoms, 117 - 122
    freedom from tyranny, 118 - 121
    of appointment viewing, 118, 119
    of limited viewing opportunities, 119, 120
    of television packages, 120, 121
    of television schedules, 119
    of the DVR, 120
    freedom of personal choice, 117, 118
    freedom of preferred advertising, 121
    freedom to watch everywhere, 121, 122
virtual server, 41
Vudu, 88
Wall Street Journal, 135
web services, 61 - 63
WebEx, 40, 41
Williams, Charlie, 28, 29
    security concerns, 28, 29
windowing, 74, 75

workday, 40
Xbox, 68
Xfinity On Demand, 47
XML, 62
YouTube, 45, 46, 58, 82, 87, 88, 110

www.ingramcontent.com/pod-product-compliance
Lightning Source LLC
Chambersburg PA
CBHW071800200526
45167CB00017B/526